Machine Learning for Time-Series with Python

Forecast, predict, and detect anomalies with
state-of-the-art machine learning methods

Ben Auffarth

BIRMINGHAM—MUMBAI

Machine Learning for Time-Series with Python

Producer: Dr. Shailesh Jain
Acquisition Editor – Peer Reviews: Saby Dsilva
Project Editor: Namrata Katare
Content Development Editor: Alex Patterson
Copy Editor: Safis Editor
Technical Editor: Aditya Sawant
Proofreader: Safis Editor
Indexer: Sejal Dsilva
Presentation Designer: Pranit Padwal

First published: October 2021

Production reference: 3110322

Published by Packt Publishing Ltd.
Livery Place
35 Livery Street
Birmingham
B3 2PB, UK.

ISBN 978-1-80181-962-6

www.packt.com

Contributors

About the author

Ben Auffarth is the author of *Artificial Intelligence with Python Cookbook*, and he co-founded and is the former president of Data Science Speakers, London. With a Ph.D. in computer science, Ben Auffarth has analyzed experiments with terabytes of data, run brain models on up to 64k cores, built systems processing hundreds of thousands of transactions per day, and trained neural networks on millions of text documents. He often encounters time-series problems in his work.

My partner was working hard over the weekends so I could concentrate, and my son of two and a half years would often tell me to get to work ("work, papa"). I'm reading lots of stories to him to make up for this time. I'd like to thank the technical reviewer for his fantastic suggestions and spotting many errors (any remaining ones are on me).

About the reviewers

Kevin Sheppard is an academic economist who specializes in the application of statistical methodology to measuring economic phenomena. His research focuses on developing statistical methodology for measuring, modeling, and forecasting measures of risk. Kevin's research is widely used in portfolio management and risk measurement. He is the maintainer of the arch and linearmodels Python packages. He is also a core contributor to statsmodels and a committer to pandas and PyData.

In 2019, his contributions to NumPy were recognized by an award from NumFocus. He has worked at the University of Oxford for the past 15 years. During this period, he has also worked for the Office of Financial Research in the U.S. Department of Treasury and has worked as a consultant to other governments and in the finance industry. Prior to joining Oxford, Kevin completed his PhD at the University of California-San Diego.

Dr Andrey Kostenko recently assumed the role of lead data scientist at the Hydroinformatics Institute (H2i.sg), a specialized consultancy and solution services provider for all aspects of water management. Prior to joining H2i, Andrey had worked as a senior data scientist at IAG InsurTech Innovation Hub for over 3 years. Before moving to Singapore in 2018, he worked as a data scientist at TrafficGuard.ai, an Australian AdTech start-up developing novel data-driven algorithms for mobile ad fraud detection. In 2013, Andrey received his doctorate degree in mathematics and statistics from Monash University, Australia, after earning an MBA degree from the UK and his first university degree from Russia.

Andrey is an enthusiastic, self-motivated, and result-oriented data science and machine learning professional, with extensive experience across a variety of disciplines and industries, including hands-on coding in R and Python to build, train, and serve time-series models for forecasting and other applications. He believes that lifelong learning and open source software are both critical for innovation in advanced analytics and artificial intelligence. Andrey is very passionate about data science in general and sequential data in particular, so one of his current focuses is on applications of deep learning to spatiotemporal data in the context of weather-related decision making.

In his spare time, Andrey is often found engaged in competitive data science projects, learning new tools across the R and Python ecosystems, exploring the latest trends in web development, solving chess puzzles, or reading about the history of science and mathematics.

Table of Contents

Preface

Time-series are ubiquitous in industry and in research. Examples of time-series can be found in healthcare, energy, finance, user behavior, and website metrics to name just a few. Due to their prevalence, time-series modeling and forecasting is crucial and it's of great economic importance to be able to model them accurately.

While traditional and well-established approaches have been dominating econometrics research and – until recently – industry, machine learning for time-series is a relatively new research field that's only recently come out of its infancy.

In the last few years, a lot of progress has been made in machine learning on time-series; however, little of this has been made available in book form for a technical audience. Many books focus on traditional techniques, but hardly deal with recent machine learning techniques. This book aims to fill this gap and covers a lot of the latest progress, as evident in results from competition such as M4, or the current state-of-the-art in time-series classification.

If you read this book, you'll learn about established as well as cutting edge techniques and tools in Python for machine learning with time-series. Each chapter covers a different topic, such as anomaly detection, probabilistic models, drift detection and adaptive online learning, deep learning models, and reinforcement learning. Each of these topics comes with a review of the latest research and an introduction to popular libraries with examples.

Who this book is for

If you want to build models that are reactive to the latest trends, seasonality, and business cycles, this is the book for you. This book is for data scientists, analysts, or programmers who want to learn more about time-series, and want to catch up on different techniques in machine learning.

What this book covers

Chapter 1, Introduction to Time-Series with Python, is a general introduction to the topic. You'll learn about time-series and why they are important, and many conventions, and you'll see an overview of applications and techniques that will be explained in more detail in dedicated chapters.

Chapter 2, Time-Series Analysis with Python, breaks down the steps for analyzing time-series. It explains statistical tests and visualizations relevant for making sense of and drawing insights from time-series.

Chapter 3, Preprocessing Time-Series, is about data treatment for time-series for traditional techniques and for machine learning. Methods such as naïve and Loess STL decomposition for seasonal and trend effects are covered, along with normalizations for values, as well as specific feature extraction techniques such as catch22 and ROCKET.

Chapter 4, Introduction to Machine Learning for Time-Series, deals with an overview of the state of the art for univariate and multivariate time-series forecasts and predictions.

Chapter 5, Forecasting with Moving Averages and Autoregressive Models, focuses on forecasting, mostly on univariate time-series (see *Chapter 12, Multivariate Forecasting* for multivariate time-series). Well-established traditional methods used in econometrics are introduced, explained, and applied on data sets.

Chapter 6, Unsupervised Methods for Time-Series, introduces anomaly detection, change detection, and clustering. The chapter reviews industry practices at major technology companies such as Facebook, Amazon, Google, and others, and gives practical examples for both anomaly detection and change detection.

Chapter 7, Machine Learning Models for Time-Series, reviews recent research on machine learning for time-series at institutes such as at the University of East Anglia and Monash University. Many techniques are summarized and compared throughout the chapter, and there's a practical section with many examples.

Chapter 8, Online Learning for Time-Series, introduces online learning, a topic often neglected. Online models continuously update their parameters based on latest samples, and some of them have mechanisms to deal with different kinds of drift – a common problem with time-series.

Chapter 9, Probabilistic Models for Time-Series, covers probabilistic models for time-series. This includes models with confidence intervals such as Facebook's Prophet, Markov Models, Fuzzy Models, and counter-factual causal models such as Bayesian Structural Time-Series Models as proposed by Google.

Chapter 10, Deep Learning for Time-Series, reviews recent literature and benchmarks for different tasks. The chapter explains techniques such as autoencoders, InceptionTime, DeepAR, N-BEATS, Recurrent Neural Networks, ConvNets, and Informer. Deep learning still hasn't completely caught up with more traditional or other machine learning techniques; however, the progress has been promising, and for certain applications such as multivariate predictions, deep learning techniques are emerging as the state of the art, as can be seen in competitions such as M4.

Chapter 11, Reinforcement Learning for Time-Series, gives an overview of basic concepts in reinforcement learning. It introduces techniques relevant for time-series such as bandit algorithms and Deep Q-Learning, and they are applied for a recommender system and for a trading algorithm.

Chapter 12, Multivariate Forecasting, gives practical examples for multivariate multistep forecasts of energy demand with deep learning models.

To get the most out of this book

- You should have a basic knowledge of Python to get started.
- All notebooks used in this book come with links to Google Colab, where you should be able to execute them.

Download the example code files

The code bundle for the book is hosted on GitHub at https://github.com/PacktPublishing/Machine-Learning-for-Time-Series-with-Python. We also have other code bundles from our rich catalog of books and videos available at https://github.com/PacktPublishing/. Check them out!

Download the color images

We also provide a PDF file that has color images of the screenshots/diagrams used in this book. You can download it here: https://static.packt-cdn.com/downloads/9781801819626_ColorImages.pdf.

Conventions used

There are a number of text conventions used throughout this book.

CodeInText: Indicates code words in text, database table names, folder names, filenames, file extensions, pathnames, dummy URLs, user input, and Twitter handles. For example; "Let's use UCBRegressor to select the best learning rate for a linear regression model."

A block of code is set as follows:

```
import numpy as np
import pandas as pd
from keras.layers import Conv1D, Input, Add, Activation, Dropout
from keras.models import Sequential, Mode
```

When we wish to draw your attention to a particular part of a code block, the relevant lines or items are set in bold:

```
owid_covid["date"] = pd.to_datetime(owid_covid["date"]
```

Any command-line input or output is written as follows:

```
pip install xgboost
```

Bold: Indicates a new term, an important word, or words that you see on the screen, for example, in menus or dialog boxes, also appear in the text like this. For example: "The task of identifying, quantifying, and decomposing these and other characteristics is called **time-series analysis**."

> Warnings or important notes appear like this.

> Tips and tricks appear like this.

Get in touch

Feedback from our readers is always welcome.

General feedback: Email feedback@packtpub.com, and mention the book's title in the subject of your message. If you have questions about any aspect of this book, please email us at questions@packtpub.com.

Errata: Although we have taken every care to ensure the accuracy of our content, mistakes do happen. If you have found a mistake in this book we would be grateful if you would report this to us. Please visit, `http://www.packtpub.com/submit-errata`, selecting your book, clicking on the Errata Submission Form link, and entering the details.

Piracy: If you come across any illegal copies of our works in any form on the Internet, we would be grateful if you would provide us with the location address or website name. Please contact us at `copyright@packtpub.com` with a link to the material.

If you are interested in becoming an author: If there is a topic that you have expertise in and you are interested in either writing or contributing to a book, please visit `http://authors.packtpub.com`.

Share your thoughts

Once you've read *Machine Learning for Time-Series with Python*, we'd love to hear your thoughts! Scan the QR code below to go straight to the Amazon review page for this book and share your feedback.

https://packt.link/r/1801819629

Your review is important to us and the tech community and will help us make sure we're delivering excellent quality content.

1

Introduction to Time-Series with Python

This book is about machine learning for time-series with Python, and you can see this chapter as a 101 class for time-series. In this chapter, we'll introduce time-series, the history of research into time-series, and how to use Python for time-series.

We'll start with what a time-series is and its main properties. We'll then look at the history of the study of time-series in different scientific disciplines foundational to the field, such as demography, astronomy, medicine, and economics.

Then, we'll go over the capabilities of Python for time-series and why Python is the go-to language for doing machine learning with time-series. Finally, I will describe how to install the most prominent libraries in Python for time-series analysis and machine learning, and we'll cover the basics of Python as relevant to time-series and machine learning.

We're going to cover the following topics:

- What Is a Time-Series?
    - Characteristics of Time-Series
- Time-Series and Forecasting – Past and Present
    - Demography
    - Genetics
    - Astronomy
    - Economics

- • Meteorology
- • Medicine
- • Applied Statistics
- • Python for Time-Series

But what is a time-series? Let's start with a definition!

What Is a Time-Series?

Since this is a book about time-series data, we should start with a clarification of what we are talking about. In this section, we'll introduce time-series and their characteristics, and we'll go through different kinds of problems and types of analyses relevant to machine learning and statistics.

Many disciplines, such as finance, public administration, energy, retail, and healthcare, are dominated by time-series data. Large areas of micro- and macro-economics rely on applied statistics with an emphasis on time-series analyses and modeling. The following are examples of time-series data:

- • Daily closing values of a stock index
- • Number of weekly infections of a disease
- • Weekly series of train accidents
- • Rainfall per day
- • Sensor data such as temperature measurements per hour
- • Population growth per year
- • Quarterly earnings of a company over a number of years

This is only to name but a few. Any data that deals with changes over time is a time-series.

It might be worth defining briefly what is considered a time-series.

Definition: Time-Series are datasets where observations are arranged in chronological order.

This is a very broad definition. Alternatively, we could have said that a time-series is a sequence of data points taken sequentially over time, or that a time-series is the result of a stochastic process.

Formally, we can define a time-series in two ways. The first one is as a mapping from the time domain to the domain of real numbers:

$$x: T \to \mathbb{R}^k,$$

where $T \subseteq \mathbb{R}$ and $k \in \mathbb{N}$.

Another way to define a time-series is as a stochastic process:

$$\{X_t\}_{t \in T}$$

Here, $X(t)$ or X_t denotes the value of the random variable X at time point t.

If T is a set of real numbers, it's a continuous-time stochastic process. If T is a set of integers, we call it a stochastic process in discrete time. The convention in the latter case is to write $\{X_n\}$.

Since time is the primary index of the dataset, by implication, time-series datasets describe how the world changes over time. They often deal with the question of how the past influences the presence or future.

The increase of monitoring and data collection brings with it the need for both statistical and machine learning techniques applied to time-series to predict and characterize the behavior of complex systems or components within a system. An important part of working with time-series is the question of how the future can be predicted based on the past. This is called forecasting.

Some methods allow adding business cycles as additional features. These additional features are called **exogenous** features - they are time-dependent, explanatory variables. We'll go through examples of feature generation in *chapter 3, Preprocessing Time-Series*.

Characteristics of Time-Series

Here's an extract of a time-series dataset as an example, exported from Google Trends, on searches for Python, R, and Julia:

Month	Python: (Worldwide)	R: (Worldwide)	Julia: (Worldwide)
2004-01-01	31	13	1
2004-02-01	29	13	1
2004-03-01	33	13	1
2004-04-01	31	14	1
2004-05-01	32	13	1

Figure 1.1: Extract of a time-series dataset

This is a **multivariate** time-series, with columns for Python, R, and Julia. The first column is the index, a date column, and its period is the month. In cases, where we have only a single variable, we speak of a **univariate** series. This dataset would be univariate if we had only one programming language instead of three.

Time-Series mostly come as discrete-time, where the time difference between each point is the same. The most important characteristics of time-series are the following:

- Long-term movements of the values (**trend**)
- Seasonal variations (**seasonality**)
- Irregular or cyclic components

A trend is the general direction in which something is developing or changing, such as a long-term increase or decrease in a sequence. An example of where a trend can be observed would be global warming, the process by which the temperatures on our planet have been rising over the last half-century.

Here's a plot of global surface temperature changes over the last 100 years from the GISS Surface Temperature Analysis dataset released by NASA:

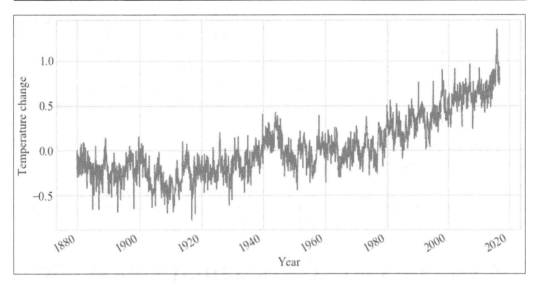

Figure 1.2: GISS surface temperature analysis from 1880 to 2019

As you can see in *Figure 1.2*, temperature changes have been varying around 0 until the mid-20th century; however, since then, there's been a clearly visible trend of an overall rise in the yearly temperature.

Seasonality is a variation that occurs at specific regular intervals of less than a year. Seasonality can occur on different time spans, such as daily, weekly, monthly, or yearly. An example of weekly seasonality would be sales of ice cream picking up each weekend. Also, depending on where you live, ice cream might only be sold in spring and summer. This is a yearly variation.

Other than seasonal changes and trends, there is variability that's not of a fixed frequency or that rises and falls in a way that's not based on seasonal frequency. Some of these we might be able to explain based on the knowledge we have.

As an example of cyclic variability that's irregular, bank holidays can fall on different calendar days each year, and promotional campaigns could depend on business decisions, such as the introduction of a new product. As an example of cyclic changes that are not seasonal, changes at the scale of milliseconds or that take place over time periods longer than a year would not be called seasonal effects.

Stationarity is the property of a time-series not to change its distribution over time as described by its summary statistics. If a time-series is stationary, it means that it has no trend and no deterministic seasonal variability, although other cyclical variability is permitted. This is an important feature for the algorithms that we'll discuss in *Chapter 5, Forecasting with Moving Averages and Autoregressive Models*. To apply them, we'll need to transform non-stationary data into stationary data by removing seasonality and trend.

We'll discuss these and other concepts in more detail in *Chapter 2, Time-Series Analysis with Python*, and *Chapter 3, Preprocessing Time-Series*.

The task of identifying, quantifying, and decomposing these and other characteristics is called **time-series analysis**. Exploratory time-series analysis is often the first step before any feature transformation and machine learning.

Time-Series and Forecasting – Past and Present

Time-Series have been studied since antiquity, and since then, time-series analysis and forecasting have come a long way. A variety of disciplines contributed to the development of techniques applied to time-series, including mathematics, astronomy, demographics, and statistics. Many innovations came initially from mathematics, later statistics, and finally machine learning. Many innovations in applied statistics had their origins in demography (used in public administration), economics, or other fields.

In this section, I'll sketch the development path from simpler methods leading up to the machine learning methods available today. I'll try to chart the development of concepts relevant to time-series from the time of the Industrial Revolution to modernity. We'll deal with the more technical and up-to-date side of things in *Chapter 4, Introduction to Machine Learning with Time-Series*.

There's still much more to come for time-series. The development of wearable sensors and the Internet of Things means that big data is available to be analyzed and used for forecasting. The availability of large datasets for benchmarks and competitions has been helping create new methods in recent years as we'll discuss in later chapters.

Demography

Much of the early work that went into establishing the theory and practice of time-series analysis came from demography as used in public administration. Many of the people mentioned in this section either worked as public servants or contributed in a private capacity out of interest in abstract problems.

John Graunt, originally a haberdasher by profession, became interested in death records as recorded by London parishes. In 1662, he published public health statistics in his book *"Natural and Political Observations Made upon the Bills of Mortality."* Among statistics about epidemiology, it included the first life table. A **life table** (also called a mortality table or actuarial table) is a table that shows, for each age, what the probability is that a person of that age will die before their next birthday. Graunt made his inferences from bills of mortality, generated by parish clerks who recorded burials in Church of England churchyards in the City of London and areas outside the city.

Graunt's book was highly influential, and he is widely regarded as the founder of demography. Graunt was elected as a fellow of the Royal Society; however, he suffered bankruptcy after his house burned down during the Great Fire of London in 1666, and he died of jaundice and liver disease at the age of 53.

Among other things, he inspired the work of Swiss mathematician Jakob Bernoulli, *"Ars Conjectandi,"* written between 1684 and 1689 and published posthumously in 1713, a landmark publication in combinatorics and probability theory that included – among many other things – a first version of the law of large numbers.

The law of large numbers describes what happens when an experiment is repeated a large number of times. Bernoulli proved that in a game of chance with two outcomes (such as a coin toss), a win or a loss, if it's repeated many times, the fraction of times that the game would be won approaches the true, expected probability.

Another major milestone in the history of demography, the statistical study of human populations, came in 1689 as an article written by Caspar Neumann, a German professor and clergyman – *"Reflexionen über Leben und Tod bei denen in Breslau Geborenen und Gestorbenen"* (translated: Reflections about the Life and Death of People Who Were Born and Died in Breslau). Neumann sent this treatise to Gottfried Leibniz, the eminent philosopher and mathematician, and later made his data available to the Royal Society in London.

Many subsequent works were based on the data and statistics in this article. In 1693, in an article on life annuities (*"An Estimate of the Degrees of the Mortality of Mankind"*) published in the *Philosophical Transactions of the Royal Society*, Edmond Halley prepared mortality tables based on Neumann's data. **Annuities** are payments made at equal intervals, such as mortgage, insurance, and pension payments. Halley's article guided the development of actuarial science and informed the British government when it came to selling retirement income insurance at an appropriate price based on the age of the purchaser. We'll encounter Halley again in the astronomy section.

Abraham de Moivre was a Frenchman who moved to England at a young age due to the religious persecution of the Huguenots in France. Today, he is best known for his work on the normal distribution and probability theory. In 1724, he published a book called "*Annuities upon Lives*," the cover of which you can see below, about mortality statistics and the foundation of the theory of annuities.

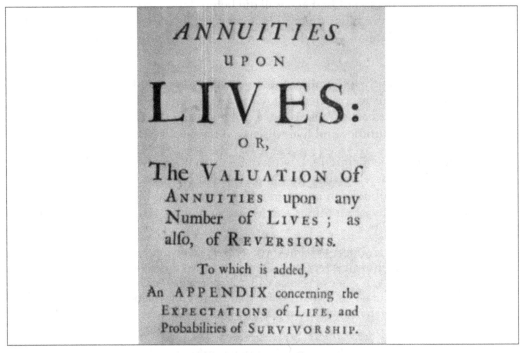

Figure 1.3: Annuities upon Lives

De Moivre is also remembered for an approximation to the binomial distribution and for his work on the Poisson distribution (later named after Siméon Denis Poisson).

With some statistics foundations in place, we are now getting into projections into the future, and this is where time-series forecasts come in. In 1751, Benjamin Franklin examined population growth and its limits in his essay "*Observations Concerning the Increase of Mankind, Peopling of Countries, etc.,*" projecting exponential growth in the British colonies. He projected a doubling of the population in the British Crown Colonies every 25 years, with the potential, he argued, to spread liberal political tradition and increase the power of England. His projection proved correct, and the exponential growth held up until the 1850s when the population of the United States surpassed England's.

Influenced by Franklin was the English cleric Thomas Robert Malthus, who feared that population growth would outstrip growth in food production. In his scenario, while population growth is exponential, the growth of food supply and other resources is linear, which would eventually lead to a collapse of society and massive population death. Writing at the end of the 18th century, he described ever-increasing famine and poverty (referred to after him as the "*Malthusian catastrophe*").

Many other statistical and mathematical concepts were worked out based on demographics data. Adolphe Quetelet, an astronomer, mathematician, and sociologist from Ghent, in today's Belgium, introduced statistical methods to social sciences to describe relationships underlying crime rates, marriage rates, and suicide rates. He called for a "social physics" that would find the laws underlying social phenomena, thus revealing the work of God. Among other things, he developed the body mass index, originally called the Quetelet index. In his 1835 book, called "*Treatise on Man*" in the English translation, he describes the concept of the average man based on normal distribution. One of Quetelet's students, Pierre Verhulst, developed the logistic function as a model of population growth.

Siméon Denis Poisson published "*Recherches sur la probabilité des jugements en matière criminelle et en matière civile*" (translated: Studies on the Probability of Judgments in Criminal and Civil Matters) in 1837, where he elaborated on probability theory for discrete occurrences that take place within a given interval. The Poisson distribution was named after him.

Wilhelm Lexis, a pioneer of the analysis of demographic time-series, published a paper called "*On the Theory of the Stability of Statistical Series*" (1879), which introduced the quantity now called the Lexis ratio. The ratio distinguishes between stable series, where underlying probability distributions giving rise to the observed rates remain constant, and non-stable series. These stable time-series would not be influenced by forces other than random noise. In today's terminology, a stable time-series would be referred to as a white noise process or a zero-order moving average.

In order to distinguish between stable and non-stable time-series, Lexis created a test statistic equal to the ratio between the dispersion of the observed rates and the dispersion that would be expected if the underlying probabilities for each of the observed rates were all equal across all of the observations. If this ratio, Q, was more than 1.41, he argued, this means the time-series is non-stable or – in his words – influenced by physical forces. Lexis later became a member of the Insurance Advisory Council for Germany's Federal Insurance Supervisory Office.

Genetics

Francis Galton, a Victorian-era English scientist, was born into an illustrious family of bankers and gun manufacturers that included several members of the Royal Society. Galton was a highly prolific writer and researcher. Today, he is mostly remembered for coining the word eugenics, the study of changes to the racial quality of future generations with a focus on desirable human qualities. Eugenics is associated with racism and white supremacy.

Galton was interested in many scientific disciplines, such as psychology, statistics, psychophysics, photography, and others, and for his contributions, he was knighted in 1909. Among other things, he contributed to anthropometry, the systematic measurement and description of human bodies. For this work, he rediscovered the concept of correlation (first developed by French physicist Auguste Bravais in 1846) and described correlations between forearm length and height, head width and head breadth, and head length and height (1888).

One of his protégés (and biographers) was Karl Pearson, born in Islington, London, into a Quaker family to a father who was Queen's Counsel (a lawyer). After studying mathematics at King's College, Cambridge, physics and philosophy at the University of Heidelberg, and physiology at the University of Berlin, he returned to London to study law. In London, he was introduced to Galton, and the two stayed in contact.

After Galton's death, Pearson was the first to hold the Chair in Eugenics endowed by Galton in his will. Pearson's main interest was in applying biometrics in the context of inheritance. He is credited with the invention of the standard deviation, a measure of the variability of the normal distribution, which replaced Carl Friedrich Gauss' concept of the mean error. He also developed contributions to statistics, including the chi-squared test, the p-value for statistical significance, correlation as it's used today, principal component analysis, and the histogram.

Pearson was succeeded as the Galton Professor of Eugenics (later renamed the Galton Chair of Genetics) by Ronald Fisher. Fisher made many innovations in evolutionary theory about mimicry, parental investment, and the Fisher principle behind the 1:1 sex ratio. In statistics, he described the linear discriminant analysis, Fisher information, the F-distribution, and the Student's t-distribution.

His contributions to statistics laid the groundwork for statistical testing in time-series analysis and some of the classical models. Fisher was made a Knight Bachelor by Queen Elizabeth II in 1952. However, his connection to racist views – for example, he endorsed the German Nazi party's policy of extermination with the goal of improving the genetic stock – has led to a recent reappraisal of his work.

As a consequence, the Ronald Fisher Centre at **University College London (UCL)** was renamed to the Centre for Computational Biology, and UCL released a public apology for its role in propagating eugenics.

Astronomy

Observations of comets and asteroids, and the and the movements of the sun and the planets have been recorded for a long time, and people have been studying these records to understand the regularities and relationships of these movements and our place in the universe. Edmond Halley, English astronomer and geophysicist, who we mentioned in the section on demography, applied Isaac Newton's laws of motion (from 1687) to comet sightings throughout history. A comet visible to the naked eye from Earth, it has been seen around the world and inscribed by astronomers and philosophers for at least about 2,000 years, appearing in Ancient Greek writings and Babylonian tables.

For instance, its appearance in 12 BCE, close to the assigned date of the birth of Jesus Christ, has led to suggestions that it might be behind the biblical story of the Star of Bethlehem. In 1066, the comet was seen in England and thought to be a divine message, a bad omen foretelling Harold II's fate when he died the same year at the Battle of Hastings fighting against Norman invaders led by William the Conqueror.

Halley connected many of these sightings and concluded that it was the same comet each time and calculated a periodicity of 75-76 years. Today it is named after him in his honor, **Halley's Comet**. This was published in the *"Synopsis of the Astronomy of Comets* (1705)"*. Halley's Comet will re-appear in 2061.

This figure shows the orbit of Halley's Comet (source: Wikimedia Commons):

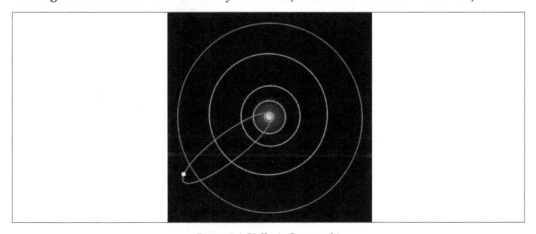

Figure 1.4: Halley's Comet orbit

German polymath Carl Friedrich Gauss devised a method for determining the orbit of the dwarf planet Ceres in 1801. Ceres orbits in the asteroid belt between Mars and Jupiter. Gauss did this based on observations of a Catholic priest and astronomer, Giuseppe Piazzi, who traced an object between January and February of the same year, before losing sight of it.

Later, line fitting was applied to the movements of celestial bodies, most prominently, the **least squares method**. It was first described by Adrien-Marie Legendre in 1805 ("*méthode des moindres carrés*"), but today it is co-credited to Gauss. Gauss published about the method later, in 1809; however, he expanded significantly beyond Legendre's work, among other things inventing the distribution named after him, the Gaussian distribution (also called the normal or Bell distribution).

 The **least squares** method is the underpinning of linear regression methods, where the parameters in sets of equations are estimated. It's a statistical procedure to find the best fit for a set of data points by minimizing the sum of the squared residuals from the plotted curve.

Only a year later, Pierre-Simon Laplace proved the **central limit theorem**, which roughly states that the sum of independent variables, even if they are not from the normal distribution, tends toward a normal distribution. This gave an important justification for the method of least squares and the normal distribution in the case of large datasets. The normal distribution has been highly influential in the field of statistics ever since, and measures such as the mean and the standard deviation are used to describe it.

Laplace was highly interested in planetary motion, but he also came up with a dynamic systems theory of tidal movements and probability theory. A fun fact to know about Laplace is that he was Napoleon Bonapart's examiner in 1784 when the latter attended the École Militaire in Paris.

One of Laplace's most famous contributions is the rule of succession, which describes the probability that an event will occur given past events. The sunrise example that he came up with for illustration of the rule of succession is the probability of the sun rising tomorrow given the number of days it has risen in the past:

$$P(s) = \frac{d+1}{d+2}$$

Laplace's assumption for the sunrise problem was that we have no knowledge of the matter other than the number of days of observations used in the formula. He actually cautioned that the application in this context is a misapplication of the rule given that we know much more about the movements of the sun and Earth.

Motivated by astronomic calculations, Augustin-Louis Cauchy invented the gradient descent optimization algorithm in 1847 (in the journal of the French Academy of Sciences, *Comptes rendus de l'Académie des Sciences*), where repeated steps in the opposite direction of the gradient led to finding a local minimum.

Many other optimization and curve-fitting innovations followed. First published in 1944 by Kenneth Levenberg and rediscovered in 1963 by Donald Marquardt, the Levenberg–Marquardt algorithm (also called the damped least-squares method) can be used for curve fitting in problems, where the dependent variables are a non-linear combination of the model parameters (non-linear problems). It combines the Gauss-Newton algorithm, a variation of the Newton algorithm published by Gauss in 1809, and the method of gradient descent invented about a hundred years earlier.

Economics

William Playfair was born in Scotland in 1759, the fourth son of a reverend's family. He took an apprenticeship with Andrew Meikle, the inventor of the threshing machine, and went on to become the personal assistant to James Watt at the Boulton and Watt steam engine manufactory.

His life was so eventful that several novels could be written about it. In 1789, he took part in the storming of the Bastille in Paris. After that he was involved as an agent of William Duer, a speculator, and the Scioto Company, in what could have been an embezzlement scheme, selling worthless deeds for land in Ohio to Frenchmen willing to emigrate. Back in London, he opened a bank that went bankrupt. Later, he was imprisoned for a few years in a debtor's prison, Fleet Prison, for being indebted. Then he went on to become a British secret agent, counterfeiting the French currency from 1789 to 1796, the assignat, to undermine the French government. The assignat soon became worthless, and inflation undermined the French government. He also patented several inventions for metalworking machinery and ships.

One of Playfair's principal achievements, however, was his popularization of several kinds of visualizations, such as the pie chart, the bar chart, and the time-series chart. He is sometimes credited with the invention of the bar chart, although Nicole Oresme showed a bar chart in a publication several hundred years earlier.

Here are two plots, the bar chart and the time-series plot, both from his "*Commercial and Political Atlas*" in 1786 (image source: Wikipedia):

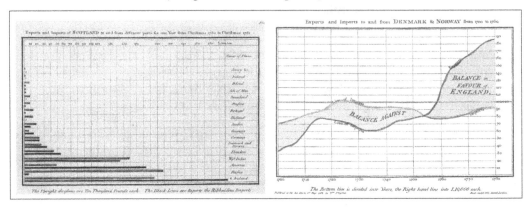

Figure 1.5: Playfair's visualizations from Commercial and Political Atlas

On the left, you can see the bar chart Playfair used for a quantitative comparison of import and export data in Scotland. On the right is the time-series chart, to show the British trade balance over time.

Meteorology

The Greek philosopher Aristotle was the first to write about weather and its measurement; however, it took much longer for the first weather predictions to be made. Vice Admiral Robert FitzRoy founded the United Kingdom's national weather service, the Meteorological Office, in 1854. FitzRoy had already reserved his place in history as the captain of the HMS Beagle, the ship that carried a recently graduated naturalist by the name of Charles Darwin around the world, playing a pivotal role in the formation of theories on evolution and natural selection.

Supported by the telegraph and the barograph, a form of barometer, the Met Office collected weather data from many different locations in London. In 1859, the steam clipper Royal Charter, while returning to Liverpool from Melbourne, Australia, shipwrecked on rocks off the Welsh coast in a storm, leading to the loss of about 450 lives. This disaster led to the development of a storm warning system that was later extended to general weather predictions. It was FitzRoy, in fact, who coined the word *forecast*, although at the time, many contemporaries referred to them as "quack weather prognostications". It is unclear whether his forecasts followed any system. He was much ridiculed by the scientific establishments for his work. He was prominently criticized by Sir Francis Galton, who had published a book called "*Meteorographica*" and later published the first weather maps. FitzRoy took his own life by cutting his throat in 1865. The storm warnings were temporarily discontinued, only to be resumed a few years later to continue to this day.

The first weather models that used atmosphere and oceans were attempted in the 1920s by Lewis Fry Richardson based on work by Norwegian Vilhelm Bjerknes, lecturer at the University of Stockholm on differential equations of fluid dynamics and thermodynamics. These models were impractical before the advent of computers, however — Richardson worked for about six weeks on a weather forecast of a limited area. His forecast turned out to be inaccurate because of numerical instability, even though his methodology was essentially correct. He abandoned his work when it became clear that his work could be of value to chemical weapons designers.

The first computerized weather models were programmed on the **Electronic Numerical Integrator and Computer** (**ENIAC**). The ENIAC, designed by John Mauchly and J. Presper Eckert, could run arbitrary sequences of operations; however, it didn't read the programs from tapes but from plugboard switches. The giant 15x9-meter machine is exhibited today at the Smithsonian Institute in Washington, D.C. Consisting of 17,500 vacuum tubes, it first produced calculations for the construction of a hydrogen bomb and was then exploited to extend forecasting past one or two days using new methods of numerical weather prediction. The computer was programmed by Klara von Neumann.

Here's a photo of the ENIAC (source: Wikimedia Commons):

Figure 1.6: Electronic Numerical Integrator and Computer (ENIAC)

You can see Betty Snyder, one of the earliest programmers of the ENIAC, standing in front of the ENIAC.

Later, Joseph Smagorinsky and Douglas Lilly developed a mathematical model for turbulence used in computational fluid dynamics. This model, the Smagorinsky-Lilly model, which is still in use today, used data about the wind, cloud cover, precipitation, atmospheric pressure, and radiation emanating from the earth and sun as input. Smagorinsky continued to lead research on global warming, investigating the climate's sensitivity to increasing carbon dioxide levels.

The introduction of mobile sensor arrays and computerized models has greatly increased the accuracy of predictions. Valuable temperature and wind data is collected by sensors deployed by meteorology offices or other sources, most importantly by commercial aircraft as they fly. Today, within a seven-day window, a forecast is accurate about 80% of the time. The grounding of commercial flights during the COVID pandemic, where there were about 75% fewer flights for some periods, has led to less accurate forecasts recently.

Medicine

In 1901, Willem Einthoven applied the string galvanometer used in the telegraph receiver to physiology. Working in Leiden, in the Netherlands, he improved upon previous designs, producing the first practical electrocardiogram (ECG). The ECG is important for monitoring and screening the function of the heart and can detect cardiac rhythm disturbances, inadequate coronary artery blood flow, and electrolyte disturbances. For the importance of this innovation, Einthoven was awarded the 1924 Nobel prize in Physiology or Medicine.

Hans Berger recorded the first human electroencephalography (EEG) recording in 1924. EEG measures the electrical activity of the brain with electrodes placed on the scalp. An EEG recording shows the brain's spontaneous electrical activity over a period of time.

Here's a graph of an EEG signal (from the EEG Eye State dataset uploaded by Oliver Roesler from DHBW, Germany):

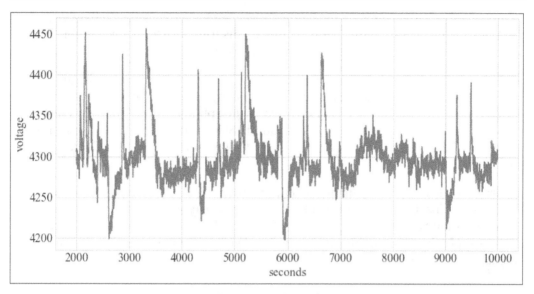

Figure 1.7: EEG signal

In EEG, the electrical activity of the brain is recorded through electrodes placed on the scalp. Its signal typically shows strong oscillations (also referred to as brain waves) at a variety of frequency ranges, most prominently these:

- Alpha (8-12 Hertz) would occur in a relaxed state, especially when closing the eyes
- Beta (16-31 Hertz) signals more active thinking
- Gamma (more than 32 Hertz) indicates cross-modal sensory processing

The medical uses of EEG are broad – among other things, EEG can be used to diagnose epilepsy, sleep disorders, tumors, stroke, depth of anesthesia, coma, and brain death.

Applied Statistics

Applied statistics and mathematics also provided inspiration and a foundation for work with time-series. Reverend Thomas Bayes (pronounced /beɪz/) proves a theorem that describes the probability of an event based on prior knowledge. Bayes' theorem is considered the foundation of Bayesian inference, a statistical inference approach.

We'll come back to this in *Chapter 9, Probabilistic Models for Time-Series*. Bayes' manuscripts were read to the Royal Society by his friend Richard Price within two years of his death (in 1761) in a heavily edited form.

The Fourier transform, which is important for filtering, converts a signal from its time domain to a representation in the frequency domain. The trigonometric decomposition of functions was discovered by Joseph Fourier in 1807, but a fast algorithm was first invented by Gauss around 1805 (although published only after his death and in Latin), and then rediscovered 160 years later by J. W. Cooley and John Tukey.

Classical time-series modeling approaches were introduced by George Box and Gwilym Jenkins in 1970 in their book *"Time-Series Analysis Forecasting and Control."* Most importantly, they formalized the ARIMA and ARMAX models and described how to apply them to time-series forecasting. We'll talk about these types of models in *Chapter 5, Forecasting with Moving Averages and Autoregressive Models*.

Python for Time-Series

For time-series, there are two main languages, R and Python, and it's worth briefly comparing the two and describing what makes Python special. Python is one of the top programming languages by popularity. According to the TIOBE from February 2021, it is only surpassed in popularity by C and Java.

Rank	Language	Ratings
1	C	16.34%
2	Java	11.29%
3	Python	10.86%
4	C++	6.88%
...	...	...
11	R	1.56%
...	...	...
29	Julia	0.52%

Figure 1.8: TIOBE language usage statistics

I've included R and Julia, two other languages used for data science, in order to support the point that Python is the most popular data science language. When comparing search volumes for Python, R, and Julia, the three foremost languages for data science, we can see that Python is much more popular than R, with Julia being the distant third. In fact, Python is ranked similar to languages such as C, Java, and C++. R is at a similar level to Assembly language and Groovy, and Julia is at the level of specialist languages such as Prolog.

R's community consists of statisticians and mathematicians, and R's strengths lie in statistics and plotting (ggplot). The weakness of R is its tooling and the virtual absence of consistent code style conventions.

On the other side, Python has been catching up in statistics and scientific computing with libraries such as NumPy, SciPy, and pandas, and it has overtaken R in both usage and usability for data science.

Python stands out in terms of machine learning libraries. The following libraries are written entirely or mainly in Python:

- Scikit-learn is written in Python and Cython (a Python dialect similar to the C programming language). It provides implementations of a very large set of algorithms for training and evaluating machine learning models.
- Statsmodels provides statistical tests, and models such as the generalized linear model (GLM), ARMA, and many more.
- Keras is an abstraction for training neural networks in Python that interact with TensorFlow and other libraries.

Some of the most popular machine learning frameworks – ones that see lots of use for development and have a large range of scalable algorithms, such as TensorFlow, PyTorch, and XGBoost – are also mainly written in Python or provide first-class interfaces for Python.

Furthermore, being a general-purpose language, Python is ideal if you want to go beyond just data analysis. With Python, you can implement the full data flow necessary for building an end-to-end machine learning system that you can deploy and integrate with the backend platforms of your company.

The following time-series plot shows the popularity of Python and R according to Google Trends. Julia is omitted because it hardly registered at the bottom of the graph.

Recently, COVID has dented the popularity of Python, but other programming languages have gone down in terms of search volumes as well.

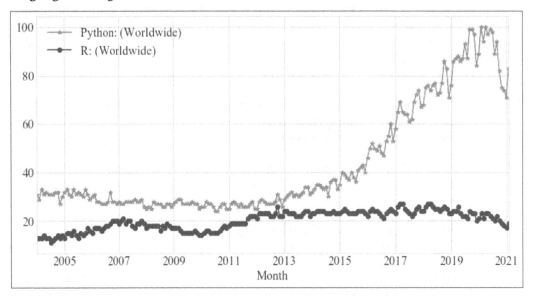

Figure 1.9: Python versus R usage over time

Python has been clearly winning out over R for the last few years, although it has to be admitted that the comparison is not completely fair since Python finds much broader application than R. However, Python is also one of the best-supported languages for data science in general and time-series in particular. As of February 2021, if we search GitHub for time-series, we find about five times the number of repositories (including repositories with Jupyter notebooks). For Julia, I found about 104 repositories. Please see the following table for the exact numbers:

Language	Repositories for time-series
Jupyter Notebook	11,297
Python	4,891
R	3,656
Julia	104

Figure 1.10: TIOBE language usage statistics

In order to just give a flavor of Python machine learning projects specializing in time-series, here's a short list of prominent libraries on GitHub:

- prophet
- sktime
- gluon-ts
- tslearn
- pyts
- seglearn
- darts
- cesium
- pmdarima

These screenshots (taken from gitcompare.com) summarize some of the statistics around these libraries, such as the number of stars (how many times someone liked the library), forks (how many times someone copied the library in order to study it or make changes), age (how long has the repository existed), and others:

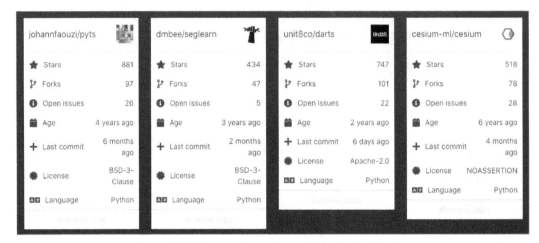

Figure 1.11: Library statistics for prominent Python libraries

We'll go through many of these time-series libraries in this book. We'll deal with a few Python data science libraries in the following sections, but if you want a full introduction to any of these libraries, you should go through a book specific to data science in Python, or even NumPy and pandas.

Installing libraries

The two main tools for maintaining and installing the Python libraries you'll need for this book are conda and pip.

Please note that the commands within the next two subsections should be executed from the system terminal or – in the case of conda – using the Anaconda navigator. For Windows and Mac users, there are graphical user interfaces available, where libraries can be searched and installed, instead of relying on the terminal.

conda works with Python, R, and other languages for the management of dependencies, and for environment encapsulation. conda helps with the installation of system libraries as well by maintaining lists of libraries associated with Python libraries.

The best way to get started with conda is to install anaconda by following the instructions from this link: https://docs.continuum.io/anaconda/install/.

There's also a graphical interface to conda that comes with a slick design, as this screenshot shows:

Figure 1.12: Anaconda navigator

The Anaconda navigator can be installed on macOS and Microsoft Windows.

Alternatively, you can rely completely on the terminal. For example, here's how to install the NumPy library from your terminal:

```
conda install numpy
```

As a side note – if you want to work with the R programming language, you can use conda, too:

```
conda install r-caret
```

See the conda documentation for an in-depth introduction and tutorials. Conda also installs versions of Python and pip, so you can use either pip or conda to install Python libraries, while having your environment managed with conda.

Terminal commands can be executed either from your system terminal or from within the Jupyter environment, the notebook or JupyterLab, by prefixing an exclamation mark.

For example, a command from within the terminal:

```
pip install xgboost
```

Can be written within the Jupyter environment as follows:

```
!pip install xgboost
```

The exclamation mark from within Jupyter tells the interpreter that this is a shell command. In recent versions of Jupyter, the exclamation mark is not necessary anymore with the pip command.

Let's take a quick look at how a simple session of starting Python and installing a library could appear on the terminal:

Figure 1.13: Terminal window

pip is a package manager for Python libraries. Here are some useful commands from your terminal:

```
# install NumPy:
pip install numpy
# install a particular version:
pip install numpy==1.20.0
# upgrade a library:
pip install -U numpy
# install all libraries listed in a requirements file:
pip install -r production/requirements.txt
# write a list of all installed libraries and their versions to a file:
pip freeze > production/requirements.txt
```

You can install different versions of Python and pip and different versions of libraries. These can be maintained as environments that you can switch between. Virtualenv is a tool to maintain environments:

```
# create a new environment myenv:
virtualenv myenv
# activate the myenv environment:
source myenv/bin/activate
# install dependencies or run python, for example:
python
# leave the environment again:
deactivate
```

The `activate` command will change your $PATH variable to point to the `virtualenv bin/` directory, which contains versions of Python and pip executables, among other things. This means you have all of those available to use as options. You should usually see the prompt reflect this change.

Please note that for the activation of the environment, you can use a complete or relative path. In Windows, the activation command is slightly different – you'd run a shell script:

```
# activate the myenv environment:
myenv\Scripts\activate.bat
```

Jupyter Notebook and JupyterLab

Jupyter stands for Julia, Python, R. It's a platform to run scripts in these and other supported languages, such as Scala and C, in an interactive environment.

You can start up a notebook server on your computer from the terminal like this:

```
jupyter notebook
```

You should see your browser opening a new tab with the Jupyter notebook. The beginning of my notebook for loading the data science language time-series looks like this:

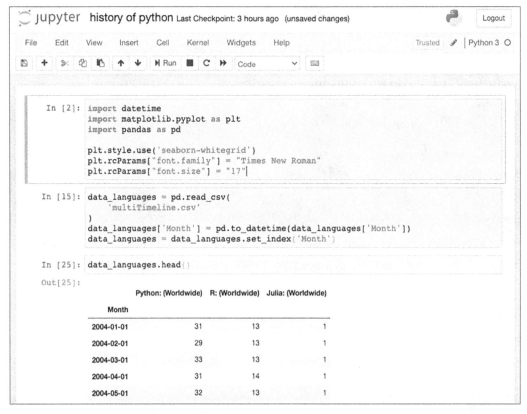

Figure 1.14: Jupyter notebook

Alternatively, we can also use JupyterLab, the next-generation notebook server that brings significant improvements in usability.

You can start up a JupyterLab notebook server from the terminal like this:

```
jupyter lab
```

JupyerLab looks a bit different from the default Jupyter server, as you can see in the screenshot below (from the JupyterLab GitHub repo):

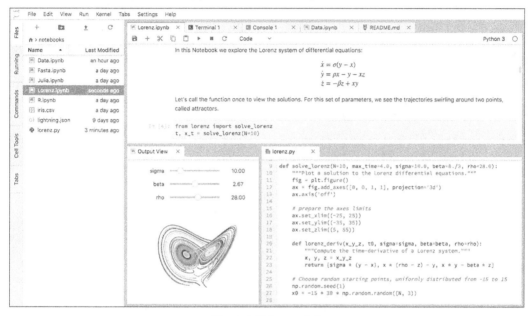

Figure 1.15: JupyterLab

Either one of these two, the Jupyter notebook or JupyterLab, will give you an **integrated development environment (IDE)** to work on the code that we'll be introducing in this book.

Finally, it's very handy to know how to get help from within Jupyter. This is where the question mark comes in. The question mark, ?, is used to provide in-notebook help like so:

```
import pandas as pd

?pd.DataFrame

Init signature:
pd.DataFrame(
    data=None,
    index: 'Optional[Axes]' = None,
    columns: 'Optional[Axes]' = None,
    dtype: 'Optional[Dtype]' = None,
    copy: 'bool' = False,
)
Docstring:
Two-dimensional, size-mutable, potentially heterogeneous tabular data.

Data structure also contains labeled axes (rows and columns).
Arithmetic operations align on both row and column labels. Can be
thought of as a dict-like container for Series objects. The primary
pandas data structure.
```

Figure 1.16: In-notebook help

You can also use single or double question marks at the end of a function if you want to access the signature or the complete source code listing of the function. This functionality can save a lot of time – instead of searching Google for the code or the definition of classes or functions, you can get to the information in milliseconds.

NumPy

NumPy is a foundational library for scientific computing in Python because so many libraries depend on it. Libraries such as PyTorch and TensorFlow provide an interface with NumPy so that data import/export is a breeze. pandas is basically a high-level interface around NumPy arrays.

SciPy also builds on top of NumPy. SciPy stands for *scientific Python* and contains functionality ranging from mathematical constants to integration, optimization, interpolation, signal processing, and more.

NumPy allows you to work with matrices in different dimensions and perform computations on them. You might work mainly with pandas or other libraries and never come much in contact with NumPy; however, for a deeper understanding and for high performance, it's definitely important to know NumPy.

A few basic commands in NumPy are below. This is supposed to be executed in the Python interpreter. We'll create one-dimensional and two-dimensional arrays:

```
import numpy as np
# 1 dimensional array:
x1 = np.array([1, 2, 3])
>>> array([0, 1, 2])

x2 = np.arange(3)
>>> array([0, 1, 2])
x1 == x2
>>> True

# 2 dimensional array:
y = np.array([(1, 2, 3),(4, 5, 6)])
```

NumPy has very handy functions for documentation; for example, to retrieve the documentation for the `optimize.fmin` function, use this (I've omitted a few lines for conciseness):

```
>> np.info(optimize.fmin)
fmin(func, x0, args=(), xtol=0.0001, ftol=0.0001, maxiter=None,
maxfun=None,full_output=0, disp=1, retall=0, callback=None)

Minimize a function using the downhill simplex algorithm.

Parameters
----------
func : callable func(x,*args)
    The objective function to be minimized.
x0 : ndarray
    Initial guess.
args : tuple
    Extra arguments passed to func, i.e. ``f(x,*args)``.
callback : callable
```

```
    Called after each iteration, as callback(xk), where xk is the
    current parameter vector.

Returns
-------
xopt : ndarray
    Parameter that minimizes function.
fopt : float
    Value of function at minimum: ``fopt = func(xopt)``.
iter : int
    Number of iterations performed.

…

Notes
-----
Uses a Nelder-Mead simplex algorithm to find the minimum of function of
one or more variables.
```

pandas

pandas is a library that allows accessing matrices or arrays as tables, by indexes such as column names – this is called a **DataFrame**. A single column or single row can be accessed as a series, another datatype in pandas. These series are NumPy arrays.

The pandas library includes functions and classes for importing and exporting data from/to CSV, Excel, and many other formats; for selecting and slicing data; and for merge, join, groupby, and aggregation functions reminiscent of Structured Query Language (SQL). You can also plot directly from pandas, because pandas is integrated with matplotlib, but it also works with other plotting libraries such as bokeh:

```python
import pandas as pd

# read a csv file:
df = pd.read_csv('value.csv')

# find how many rows in a dataframe:
len(df)
```

```python
# return the head or tail of a dataframe:
df.head()
df.tail()

# print the full dataframe:
with pd.option_context(
    'display.max_rows', None,
    'display.max_columns', None
):
    print(df)

# create a dataframe:
df2 = pd.DataFrame({"A": [1, 2], "B": [3, 4]})
# plot two columns against each other:
df2.plot(x='A', y='B', kind='scatter')

# save the dataframe to a csv:
df2.to_csv('new_file.csv', index=False)

# output to NumPy matrix:
df2.to_numpy()
```

The output of the last command should look like this:

```
array([[1, 3],
       [2, 4]])
```

Best practice in Python

In this section, I want to talk about good coding. Whole books have been written about this, and this one section cannot do justice to this matter; however, I aim to at least give some essentials and pointers. For coding beyond the beginner's level and within a corporate environment, or any organization for that matter, good practice takes on importance.

It's not easy and takes experience to write generalizable code that lends itself to maintenance and enhancements. Only code that expresses ideas in a way that is readable to other people will be useful for a team. One of the most important principles is DRY (don't repeat yourself), where repetition is reduced and each functionality finds its unique representation within the system.

This is not a full list, but some other principles include the following:

- Documentation
- Dependency management
- Code validation
- Error handling
- Testing (in particular, unit testing)

Some of these have entire books written about them, and it's outside our scope here to go into detail. Each of these is crucially important once you gear up to production so your code can be relied on.

It still happens to me that I feel like an idiot whenever I return to one of my projects, be it in my job or private life, and realize I didn't write enough documentation. When that happens, I have to expend energy rebuilding the correct mental representation of my code. If done properly, writing documentation can help you in your flow. Other people reading your code, and your future self in a few months, will be glad you wrote documentation, especially for functions, classes, and modules (docstrings).

Encapsulation of dependencies means that your code is isolated, portable, and reproducible. Two main tools have emerged during the last few years for the management of dependencies and environments in Python, conda and pip, which we've mentioned in the previous section.

A mishmash of styles and conventions render any project a mess that's not only hard to read but hard to maintain. One of the most important coding styles for Python is Python Enhancement Proposal 8, or PEP 8 for short. You can find the style guide for PEP 8 at http://bit.ly/3evsgIW.

A few tools have been developed to check Python code for adherence to PEP 8 (and a few additional conventions). These tools can help you make your code more legible and maintainable while saving time and mental energy; for example, Flake8, Black, mypy, or Pylint. Flake8 and Pylint not only check for coding style but also for common coding mistakes and potential bugs. If you want to run a Flake8 test on a Python script, you can type this, for example:

```
flake8 --shore-source --show-pep8 myscript.py
```

Black can nag you about formatting or automatically fix formats in a file, module, or even a whole project. pydocstyle checks for the existence of documentation and the compliance of the documentation with documentation style guidelines.

Further, more in-depth development and coding styles have been created by developers from several high-level projects and can be very instructive. The guide for the scikit-learn project can be found at `http://bitly.com/3etFrtz`. For pandas, you can compare the styles at `http://bit.ly/2OlpCKZ`.

Unit testing is a method by which you set up tests for modules, classes, and other units of code. One of the most popular libraries for unit testing in Python is pytest. You can find out more about pytest in the pytest documentation: `https://docs.pytest.org/en/stable/`

Summary

In this chapter, we've introduced time-series, the history of research into time-series, and Python for time-series.

We started with a definition of time-series and its main properties. We then looked at the history of the study of time-series in different scientific disciplines, such as demography and genetics, astronomy, economics, meteorology, medicine, and applied statistics.

Then, we went over the capabilities of Python for time-series and why Python is the go-to language for doing machine learning with time-series. Finally, I described how to install and use Python for time-series analysis and machine learning, and we covered some of the basics of Python as relevant to time-series and machine learning.

In the next chapter, we'll look at time-series analysis with Python.

2

Time-Series Analysis with Python

Time-Series analysis revolves around getting familiar with a dataset and coming up with ideas and hypotheses. It can be thought of as "storytelling for data scientists" and is a critical step in machine learning, because it can inform and help shape tentative conclusions to test while training a machine learning model. Roughly speaking, the main difference between time-series analysis and machine learning is that time-series analysis does not include formal statistical modeling and inference.

While it can be daunting and seem complex, it is a generally very structured process. In this chapter, we will go through the fundamentals in Python for dealing with time-series patterns. In Python, we can do time-series analysis by interactively querying our data using a number of tools that we have at our fingertips. This starts from creating and loading time-series datasets to identifying trend and seasonality. We'll outline both the structure of time-series analysis, and the constituents both in terms of theory and practice in Python by going through examples.

The main example will use a dataset of air pollution in London and Delhi. You can find this example as a Jupyter notebook in the book's GitHub repository.

We're going to cover the following topics:

- What is time-series analysis?
- Working with time-series in Python
- Understanding the variables
- Uncovering relationships between variables
- Identifying trend and seasonality

We'll start with a characterization and an attempt at a definition of time-series analysis.

What is time-series analysis?

The term **time-series analysis** (**TSA**) refers to the statistical approach to time-series or the analysis of trend and seasonality. It is often an *ad hoc* exploration and analysis that usually involves visualizing distributions, trends, cyclic patterns, and relationships between features, and between features and the target(s).

More generally, we can say TSA is roughly **exploratory data analysis** (**EDA**) that's specific to time-series data. This comparison can be misleading however since TSA can include both descriptive and exploratory elements.

Let's see quickly the differences between descriptive and exploratory analysis:

- **Descriptive analysis** summarizes characteristics of a dataset
- **Exploratory analysis** analyzes for patterns, trends, or relationships between variables

Therefore, TSA is the initial investigation of a dataset with the goal of discovering patterns, especially trend and seasonality, and obtaining initial insights, testing hypotheses, and extracting meaningful summary statistics.

 Definition: Time-Series Analysis (TSA) is the process of extracting a summary and other statistical information from time-series, most importantly, the analysis of trend and seasonality.

Since an important part of TSA is gathering statistics and representing your dataset graphically through visualization, we'll do a lot of plots in this chapter. Many statistics and plots described in this chapter are specific to TSA, so even if you are familiar with EDA, you'll find something new.

A part of TSA is collecting and reviewing data, examining the distribution of variables (and variable types), and checking for errors, outliers, and missing values. Some errors, variable types, and anomalies can be corrected, therefore EDA is often performed hand in hand with preprocessing and feature engineering, where columns and fields are selected and transformed. The whole process from data loading to machine learning is highly iterative and may involve multiple instances of TSA at different points.

Here are a few crucial steps for working with time-series:

- Importing the dataset
- Data cleaning
- Understanding variables
- Uncovering relationships between variables
- Identifying trend and seasonality
- Preprocessing (including feature engineering)
- Training a machine learning model

Importing the data can be considered prior to TSA, and data cleaning, feature engineering, and training a machine learning model are not strictly part of TSA.

Importing the data includes parsing, for example extracting dates. The three steps that are central to TSA are understanding variables, uncovering relationships between variables, and identifying trend and seasonality. There's a lot more to say about each of them, and in this chapter, we'll talk about them in more detail in their dedicated sections.

The steps belonging to TSA and leading to preprocessing (feature engineering) and machine learning are highly iterative, and can be visually appreciated in the following time-series machine learning flywheel:

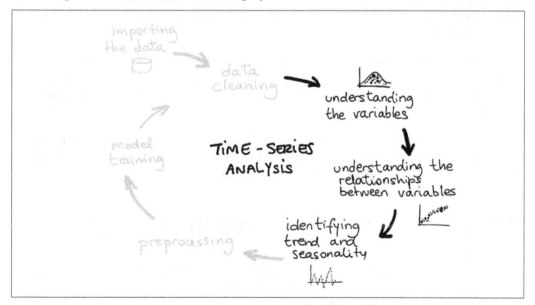

Figure 2.1: The time-series machine learning flywheel

This flywheel emphasizes the iterative nature of the work. For example, data cleaning comes often after loading the data, but will come up again after we've made another discovery about our variables. I've highlighted TSA in dark, while steps that are not strictly part of TSA are grayed out.

Let's go through something practical! We'll start by loading a dataset. Right after importing the data, we'd ask questions like what's the size of the dataset (the number of observations)? How many features or columns do we have? What are the column types?

We'll typically look at histograms or distribution plots. For assessing relationships between features and target variables, we'd calculate correlations and visualize them as a correlation heatmap, where the correlation strength between variables is mapped to colors.

We'd look for missing values – in a spreadsheet, these would be empty cells – and we'd clean up and correct these irregularities, where possible.

We are going to be analyzing relationships between variables, and in TSA, one of its peculiarities is that we need to investigate the relationship of time with each variable.

Generally, a useful way of distinguishing different types of techniques could be between univariate and multivariate analysis, and between graphical and non-graphical techniques. **Univariate analysis** means we are looking at a single variable. This means we could be inspecting values to get the means and the variance, or – for the graphical side – plotting the distribution. We summarize these techniques in the *Understanding the variables* section.

On the other hand, **multivariate analysis** means we are calculating correlations between variables, or – for the graphical side – drawing a scatter plot, for example. We'll delve into these techniques in the *Uncovering relationships between variables* section.

Before we continue, let's go through a bit of the basics of time-series with Python. This will cover the basic operations with time-series data as an introduction. After this, we'll go through Python commands with an actual dataset.

Working with time-series in Python

Python has a lot of libraries and packages for time-series, such as `datetime`, `time`, `calendar`, `dateutil`, and `pytz`, which can be highly confusing for beginners. At the same time, there are many different data types like `date`, `time`, `datetime`, `tzinfo`, `timedelta`, `relativedelta`, and more.

When it comes to using them, the devil is in the details. Just to name one example: many of these types are insensitive to the timezone. You should feel reassured, however, knowing that to get started, familiarity with a small subset of these libraries and data types is enough.

Requirements

In this chapter, we'll use several libraries, which we can quickly install from the terminal (or similarly from Anaconda Navigator):

```
pip install -U dython scipy numpy pandas seaborn scikit-learn
```

We'll execute the commands from the Python (or IPython) terminal, but equally we could execute them from a Jupyter notebook (or a different environment).

It's a good start if we at least know datetime and pandas, two very prominent libraries, which we'll cover in the following two sections. We'll create basic objects and do simple manipulations on them.

Datetime

The date and datetime data types are not primitive types in Python the way that numbers (float and int), string, list, dictionary, tuple, or file are. To work with date and datetime objects, we have to import datetime, a library that is part of the Python Standard Library, and the libraries that come by default with CPython and other main Python distributions.

datetime comes with objects such as date, datetime, time, and timedelta, among others. The difference between datetime and date objects is that the datetime object includes time information in addition to a date.

To get a date, we can do this:

```
from datetime import date
```

To get today's date:

```
today = date.today()
```

To get some other date:

```
other_date = date(2021, 3, 24)
```

If we want a `datetime` object (a timestamp) instead, we can do this as well:

```
from datetime import datetime
now = datetime.now()
```

This will get the current timestamp. We can create a `datetime` for a specific date and time as well:

```
some_date = datetime(2021, 5, 18, 15, 39, 0)
some_date.isoformat()
```

We can get a string output in isoformat:

```
'2021-05-18T15:39:00'
```

isoformat, short for the ISO 8601 format, is an international standard for representing dates and times.

We can also work with time differences using `timedelta`:

```
from datetime import timedelta
year = timedelta(days=365)
```

These `timedelta` objects can be added to other objects for calculations. We can do calculations with a `timedelta` object, for example:

```
year * 10
```

This should give us the following output:

```
datetime.timedelta(days=3650)
```

The datetime library can parse string inputs to `date` and `datetime` types and output these objects as `string`:

```
from datetime import date
some_date = date.fromisoformat('2021-03-24')
```

Or:

```
some_date = datetime.date(2021, 3, 24)
```

We can format the output with string format options, for example like this:

```
some_date.strftime('%A %d. %B %Y')
```

This would give us:

```
'Wednesday 24. March 2021'
```

Similarly, we can read in a `date` or `datetime` object from a string, and we can use the same format options:

```
from datetime import datetime
dt = datetime.strptime('24/03/21 15:48', '%d/%m/%y %H:%M')
```

You can find a complete list of formatting options that you can use both for parsing strings and printing `datetime` objects here: `https://strftime.org/`.

A few important ones are listed in this table:

Format string	Meaning
%Y	Year as 4 digits
%y	Year as 2 digits
%m	Month as a number
%d	Day
%H	Hour as 2 digits
%M	Minute as 2 digits

Figure 2.2: Format strings for dates

It's useful to remember these strings with formatting options. For example, the format string for a US date separated by slashes would look like this:

```
'%d/%m/%Y'
```

pandas

We introduced the pandas library in the previous chapter. pandas is one of the most important libraries in the Python ecosystem for data science, used for data manipulation and analysis. Initially released in 2008, it has been a major driver of Python's success.

pandas comes with significant time-series functionality such as date range generation, frequency conversion, moving window statistics, date shifting, and lagging.

Let's go through some of these basics. We can create a time-series as follows:

```
import pandas as pd
pd.date_range(start='2021-03-24', end='2021-09-01')
```

This gives us a `DateTimeIndex` like this:

```
DatetimeIndex(['2021-03-24', '2021-03-25', '2021-03-26', '2021-03-27',
               '2021-03-28', '2021-03-29', '2021-03-30', '2021-03-31',
               '2021-04-01', '2021-04-02',
               ...
               '2021-08-23', '2021-08-24', '2021-08-25', '2021-08-26',
               '2021-08-27', '2021-08-28', '2021-08-29', '2021-08-30',
               '2021-08-31', '2021-09-01'],
              dtype='datetime64[ns]', length=162, freq='D')
```

We can also create a time-series as follows:

```
pd.Series(pd.date_range("2021", freq="D", periods=3))
```

This would give us a time-series like this:

```
0    2021-01-01
1    2021-01-02
2    2021-01-03
dtype: datetime64[ns]
```

As you can see, this type is called a `DatetimeIndex`. This means we can use this data type for indexing a dataset.

One of the most important functionalities is parsing to `date` or `datetime` objects from either `string` or separate columns:

```
import pandas as pd
df = pd.DataFrame({'year': [2021, 2022],
    'month': [3, 4],
    'day': [24, 25]}
)

ts1 = pd.to_datetime(df)
ts2 = pd.to_datetime('20210324', format='%Y%m%d')
```

We've created two time-series.

You can take a rolling window for calculations like this:

```
s = pd.Series([1, 2, 3, 4, 5])
s.rolling(3).sum()
```

Can you guess the result of this? If not, why don't you put this into your Python interpreter?

A time-series would usually be an index with a time object and one or more columns with numeric or other types, such as this:

```
import numpy as np
rng = pd.date_range('2021-03-24', '2021-09-01', freq='D')
ts = pd.Series(np.random.randn(len(rng)), index=rng)
```

We can have a look at our time-series:

```
2021-03-24    -2.332713
2021-03-25     0.177074
2021-03-26    -2.136295
2021-03-27     2.992240
2021-03-28    -0.457537
                 ...
2021-08-28    -0.705022
2021-08-29     1.089697
2021-08-30     0.384947
2021-08-31     1.003391
2021-09-01    -1.021058
Freq: D, Length: 162, dtype: float64
```

We can index these time-series datasets like any other pandas Series or DataFrame. `ts[:2].index` would give us:

```
DatetimeIndex(['2021-03-24', '2021-03-25'], dtype='datetime64[ns]',
freq='D')
```

Interestingly, we can index directly with strings or datetime objects. For example, `ts['2021-03-28':'2021-03-30']` gives us:

```
2021-03-28    -0.457537
2021-03-29    -1.089423
2021-03-30    -0.708091
Freq: D, dtype: float64
```

You can shift or lag the values in a time-series back and forward in time using the `shift` method. This changes the alignment of the data:

```
ts.shift(1)[:5]
```

We can also change the resolution of time-series objects, for example like this:

```
ts.asfreq('M')
```

 Please note the difference between `datetime` and `pd.DateTimeIndex`. Even though they encode the same kind of information, they are different types and they might not always play well with each other. Therefore, I'd recommend to always explicitly convert types when doing comparisons.

In the next section, let's go through a basic example of importing a time-series dataset, getting summary statistics, and plotting some variables.

Understanding the variables

We're going to load up a time-series dataset of air pollution, then we are going to do some very basic inspection of variables.

This step is performed on each variable on its own (univariate analysis) and can include summary statistics for each of the variables, histograms, finding missing values or outliers, and testing stationarity.

The most important descriptors of continuous variables are the mean and the standard deviation. As a reminder, here are the formulas for the mean and the standard deviation. We are going to build on these formulas later with more complex formulas. The **mean** usually refers to the arithmetic mean, which is the most commonly used average and is defined as:

$$\overline{x} = \frac{1}{n}\sum_{i=1}^{n} x_i$$

The **standard deviation** is the square root of the average squared difference to this mean value:

$$\sigma_X = \sqrt{\frac{1}{n}\sum_{i=1}^{n} (x_i - \bar{x})^2} = \sqrt{\mathbb{E}[(X - \bar{x})^2}$$

The **standard error (SE)** is an approximation of the standard deviation of sampled data. It measures the dispersion of sample means around the population mean, but normalized by the root of the sample size. The more data points involved in the calculation, the smaller the standard error tends to be. The SE is equal to the standard deviation divided by the square root of the sample size:

$$\sigma_{\overline{x}} = \frac{\sigma}{\sqrt{n}}$$

An important application of the SE is the estimation of confidence intervals of the mean. A **confidence interval** gives a range of values for a parameter. For example, the 95th percentile upper confidence limit, CI^+, is defined as:

$$CI^+ = \overline{x} + (SE \times 1.96)$$

Similarly, replacing the plus with a minus, the lower confidence interval is defined as:

$$CI^- = \overline{x} - (SE \times 1.96)$$

The **median** is another average, particularly useful when the data can't be described accurately by the mean and standard deviations. This is the case when there's a long tail, several peaks, or a skew in one or the other direction. The median is defined as:

$$\text{median}(X) = x_{(n+1)/2}$$

This assumes that X is ordered by value in ascending or descending direction. Then, the value that lies in the middle, just at $(n + 1)/2$, is the median. The median is the 50th **percentile**, which means that it is higher than exactly half or 50% of the points in X. Other important percentiles are the 25th and the 75th, which are also the first **quartile** and the third quartile. The difference between these two is called the **interquartile range**.

These are the most common descriptors, but not the only ones even by a long stretch. We won't go into much more detail here, but we'll see a few more descriptors later.

Let's get our hands dirty with some code!

We'll import datetime, pandas, matplotlib, and seaborn to use them later. Matplotlib and seaborn are libraries for plotting. Here it goes:

```
import datetime
import pandas as pd
import matplotlib.pyplot as plt
```

Then we'll read in a CSV file. The data is from the **Our World in Data (OWID)** website, a collection of statistics and articles about the state of the world, maintained by Max Roser, research director in economics at the University of Oxford.

We can load local files or files on the internet. In this case, we'll load a dataset from GitHub. This is a dataset of air pollutants over time. In pandas you can pass the URL directly into the `read_csv()` method:

```python
pollution = pd.read_csv(
    'https://raw.githubusercontent.com/owid/owid-datasets/master/
datasets/Air%20pollution%20by%20city%20-%20Fouquet%20and%20DPCC%20
(2011)/Air%20pollution%20by%20city%20-%20Fouquet%20and%20DPCC%20(2011).
csv'
)
len(pollution)
```

```
331
```

```python
pollution.columns
```

```
Index(['Entity', 'Year', 'Smoke (Fouquet and DPCC (2011))',
       'Suspended Particulate Matter (SPM) (Fouquet and DPCC (2011))'],
      dtype='object')
```

If you have problems downloading the file, you can download it manually from the book's GitHub repository from the `chapter2` folder.

Now we know the size of the dataset (331 rows) and the column names. The column names are a bit long, let's simplify it by renaming them and then carry on:

```python
pollution = pollution.rename(
    columns={
        'Suspended Particulate Matter (SPM) (Fouquet and DPCC (2011))':
            'SPM',
        'Smoke (Fouquet and DPCC (2011))' : 'Smoke',
        'Entity': 'City'
    }
)
pollution.dtypes
```

Here's the output:

```
City                    object
Year                     int64
Smoke                  float64
```

```
SPM                              float64
dtype: object
```

```
pollution.City.unique()
```

```
array(['Delhi', 'London'], dtype=object)
```

```
pollution.Year.min(), pollution.Year.max()
```

The minimum and the maximum year are these:

```
(1700, 2016)
```

pandas brings lots of methods to explore and discover your dataset – `min()`, `max()`, `mean()`, `count()`, and `describe()` can all come in very handy.

City, Smoke, and SPM are much clearer names for the variables. We've learned that our dataset covers two cities, London and Delhi, and over a time period between 1700 and 2016.

We'll convert our Year column from `int64` to `datetime`. This will help with plotting:

```
pollution['Year'] = pollution['Year'].apply(
    lambda x: datetime.datetime.strptime(str(x), '%Y')
)
pollution.dtypes
```

```
City              object
Year       datetime64[ns]
Smoke             float64
SPM               float64
dtype: object
```

Year is now a `datetime64[ns]` type. It's a `datetime` of 64 bits. Each value describes a nanosecond, the default unit.

Let's check for missing values and get descriptive summary statistics of columns:

```
pollution.isnull().mean()
```

```
City              0.000000
Year              0.000000
Smoke             0.090634
SPM               0.000000
dtype: float64
```

```
pollution.describe()
```

	Smoke	SPM
count	301.000000	331.000000
mean	210.296440	365.970050
std	88.543288	172.512674
min	13.750000	15.000000
25%	168.571429	288.474026
50%	208.214286	375.324675
75%	291.818182	512.609209
max	342.857143	623.376623

The Smoke variable has 9% missing values. For now, we can just focus on the SPM variable, which doesn't have any missing values.

The pandas describe() method gives us counts of non-null values, mean and standard deviation, 25th, 50th, and 75th percentiles, and the range as the minimum and maximum.

A **histogram**, first introduced by Karl Pearson, is a count of values within a series of ranges called bins (or buckets). The variable is first divided into a series of intervals, and then all points that fall into each interval are counted (bin counts). We can present these counts visually as a barplot.

Let's plot a histogram of the SPM variable:

```
n, bins, patches = plt.hist(
    x=pollution['SPM'], bins='auto',
    alpha=0.7, rwidth=0.85
)
plt.grid(axis='y', alpha=0.75)
plt.xlabel('SPM')
plt.ylabel('Frequency')
```

This is the plot we get:

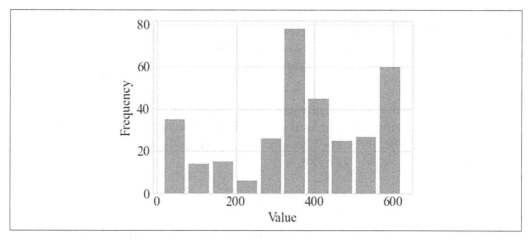

Figure 2.3: Histogram of the SPM variable

A histogram can help if you have continuous measurements and want to understand the distribution of values. Further, a histogram can indicate if there are outliers.

This closes the first part of our TSA. We'll come back to our air pollution dataset later.

Uncovering relationships between variables

If we are not dealing with a univariate time-series where there's only a single variable, the relationship between the variables needs to be investigated. This includes the direction and rough size of any correlations. This is important to avoid feature leakage and collinearity.

Feature leakage is when a variable unintentionally gives away the target. For example, the variable named amount_paid would give away the label has_paid. A more complex example would be if we were analyzing data for an online supermarket, and our dataset consisted of customer variables such as age, number of purchases in the past, length of visit, and finally the contents of their cart. What we want to predict, our target, is the result of their buying decision as either abandoned (when they canceled their purchase) or paid. We could find that a purchase is highly correlated with bags in their cart due to just the simple fact that bags are added at the last step. However, concluding we should offer bags to customers when they land on our site would probably miss the point, when it's the length of their stay that could be, in fact, the determining variable, and an intervention through a widget or customer service agent might be much more effective.

Collinearity means that independent variables (features) are correlated. The latter case can be problematic in linear models. Therefore, if we carry out linear regression and find two variables that are highly correlated between themselves, we should remove one of them or use dimensionality reduction techniques such as Principal Component Analysis (PCA).

The **Pearson correlation** coefficient was developed by Karl Pearson, whom we've discussed in the previous chapter, and named in his honor to distinguish it from other correlation coefficients. The Pearson correlation coefficient between two variables X and Y is defined as follows:

$$\rho_{X,Y} = \frac{\mathrm{cov}(X,Y)}{\sigma_X \sigma_Y}$$

$\mathrm{cov}(X,Y)$ is the covariance between the two variables defined as the expected value (the mean) between the differences of each point to the variable mean:

$$\mathrm{cov}(X,Y) = \mathbb{E}[(X - \bar{x})(Y - \bar{y})]$$

σ_X is the standard deviation of the variable X.

Expanded, the formula looks like this:

$$\hat{\rho}_{X,Y} = \frac{\sum_{i=1}^{N}(x_i - \bar{x})(y_i - \bar{y})}{\sigma_x \sigma_y}$$

There are three types of correlation: positive, negative, and no correlation. Positive correlation means that as one variable increases the other does as well. In the case of the Pearson correlation coefficient, the increase of one variable to the other should be linear.

If we looked at a plot of global life expectancy from 1800 onward, we'd see an increase of years lived with the time axis. You can see the plot of global life expectancy based on data on OWID:

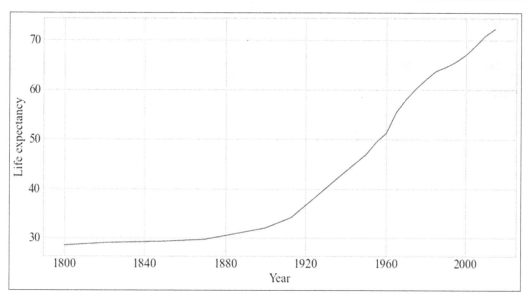

Figure 2.4: Life expectancy from 1800 to today

We can see how life expectancy has been increasing steadily since the end of the 19th century until today.

This plot is called a **run chart** or temporal **line chart**.

In order to calculate the Pearson correlation, we can use a function from SciPy:

```
from scipy import stats

def ignore_nans(a, b):
    index = ~a.isnull() & ~b.isnull()
    return a[index], b[index]

stats.pearsonr(*ignore_nans(pollution['Smoke'], pollution['SPM']))
```

Here's the Pearson correlation and the p-value that indicates significance (the lower, the more significant)

```
(0.9454809183096181, 3.313283689287137e-10
```

We see a very strong positive correlation of time with life expectancy, 0.94, at very high significance (the second number in the return). You can find more details about the dataset on the OWID website.

Conversely, we would see a negative correlation of time with child mortality – as the year increases, child mortality decreases. This plot shows the child mortality per 1,000 children on data taken from OWID:

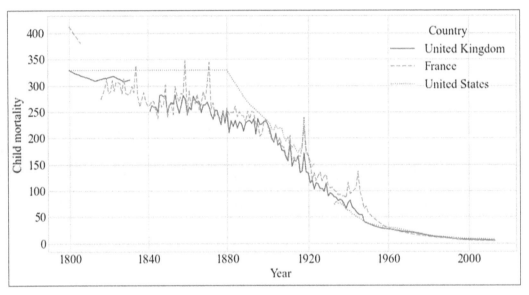

Figure 2.5: Child mortality from 1800 to today in the UK, France, and the USA

In this plot, we can see that in all three countries child mortality has been decreasing since the start of the 19th century until today.

In the case of the United States, we'll find a negative correlation of -0.95 between child mortality and time.

We can also compare the countries to each other. We can calculate correlations between each feature. In this case, each feature contains the values for the three countries.

This gives a **correlation matrix** of 3x3, which we can visualize as a heatmap:

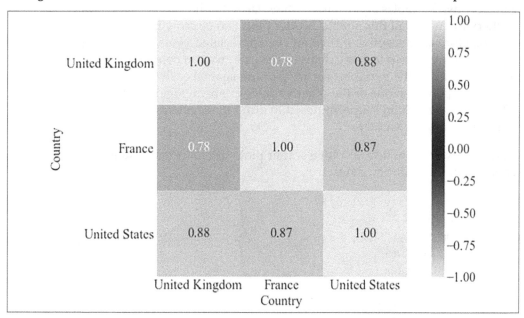

Figure 2.6: Correlation heatmap of child mortality between the UK, France, and the USA

In this correlation heatmap, we can see that countries are highly correlated (for example, a correlation of 0.78 between France and the United Kingdom).

The diagonal of the correlation matrix is always 1.0, and the matrix is symmetrical across the diagonal. Therefore, sometimes we only show the lower triangle below the diagonal (or sometimes the upper triangle). We can see that child mortality in the United Kingdom is more similar to that of the United States than that of France.

Does this mean that the UK went through a similar development as the United States? These statistics and visualizations can often generate questions to answer, or hypotheses that we can test.

As mentioned before, the full notebooks for the different datasets are available on GitHub, however, here's the snippet for the heatmap:

```
import dython
dython.nominal.associations(child_mortality[countries], figsize=(12, 6));
```

The correlation coefficient struggles with cases where the increases are non-linear or non-continuous, or (because of the squared term) when there are outliers. For example, if we looked at air pollution from the 1700s onward, we'd see a steep increase in air pollutants from coal and – with the introduction of the steam engine – a decrease in pollutants.

A **scatter plot** can be used for showing and comparing numeric values. It plots values of two variables against each other. Usually, the variables are numerical – otherwise, we'd call this a table. Scatter plots can be crowded in certain areas, and therefore are deceptive if this can't be appreciated visually. Adding jitter and transparency can help to some degree, however, we can combine a scatter plot with the histograms of the variables we are plotting against each other, so we can see how many points on one or the other variable are being displayed. Scatter plots often have a best-fit line superimposed in order to visualize how one variable is the function of another variable.

Here's an example of how to plot a scatter plot with marginal histograms of the two variables in the pollution dataset:

```python
plt.figure(figsize=(12, 6))
sns.jointplot(
    x="Smoke", y="SPM",
    edgecolor="white",
    data=pollution
)
plt.xlabel("Smoke")
plt.ylabel("SPM");
```

Here's the resulting plot:

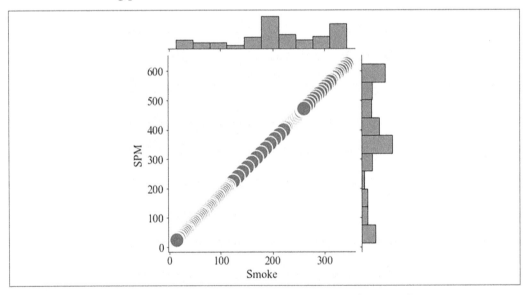

Figure 2.7: Scatter plot with marginal histograms of Smoke against SPM

In the scatter plot, we can see that the two variables are extremely similar – the values are all on the diagonal. The correlation between these two variables is perfect, 1.0, which means that they are in fact identical.

We've seen the dataset of **Suspended Particulate Matter** (**SPM**) before. Let's plot SPM over time:

```
pollution = pollution.pivot("Year", "City", "SPM")
plt.figure(figsize=(12, 6))
sns.lineplot(data=pollution)
plt.ylabel('SPM');
```

Here's the plot:

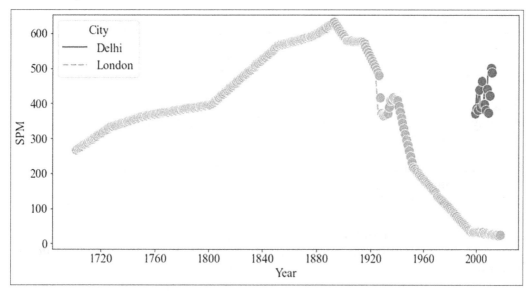

Figure 2.8: Suspended particle matter from the 1700s to today

We can see in the plot that the air quality (measured as suspended particle matter) in London was getting worse until around 1880 (presumably because of heating materials such as wood and coal), however, has since been improving.

We find a correlation coefficient of -0.36 with high significance. The steep decline of pollutants from 1880 onward dominates over the 180 years of slow growth. If we looked separately at the time from 1700 to 1880 and from 1880 to the present, we'd find 0.97 and -0.97 respectively, examples of very strong correlation and very strong anti-correlation.

The **Spearman rank correlation** can handle outliers and non-linear relationships much better than the Pearson correlation coefficient – although it can't handle non-continuous cases like the one above. The Spearman correlation is the Pearson correlation, only applied on ranks of variables' values instead of the variables' values directly. The Spearman correlation of the time-series for air pollution is -0.19, and for the two time periods before and after 1880 we get 0.99 and -0.99, respectively.

In the case of the Spearman correlation coefficient, the numerical differences are ignored – what counts is the order of the points. In this case, the order of the points within the two time periods aligns nearly perfectly.

In the next section, we'll talk about trend and seasonality.

Identifying trend and seasonality

Trend, seasonality, and cyclic variations are the most important characteristics of time-series. A **trend** is the presence of a long-term increase or decrease in the sequence. **Seasonality** is a variation that occurs at specific regular intervals of less than a year. Seasonality can occur on different time spans such as daily, weekly, monthly, or yearly. Finally, **cyclic variations** are rises and falls that are not of a fixed frequency.

An important characteristic of time-series is **stationarity**. This refers to a property of time-series not to change distribution over time, or in other words, that the process that produces the time-series doesn't change with time. Time-Series that don't change over time are called **stationary** (or **stationary processes**). Many models or measures assume stationarity and might not work properly if the data is not stationary. Therefore, with these algorithms, the time-series should be decomposed first into the main signal, and then the seasonal and trend components. In this decomposition, we would subtract the trend and seasonal components from the original time-series.

In this section, we'll first go through an example of how to estimate trend and seasonality using curve fitting. Then, we'll look at other tools that can help discover trends, seasonality, and cyclic variations. These include statistics such as autocorrelation and the augmented Dickey–Fuller test, and visualizations such as the autocorrelation plot (also: lag plot) and the periodogram.

Let's start with a hopefully clear example of how seasonality and trend can be estimated in just a few lines of Python. For this, we'll come back to the GISS Surface Temperature Analysis dataset released by NASA. We'll load the dataset, and we'll do curve fitting, which comes straight out of the box in NumPy.

We'll download the dataset from Datahub (`https://datahub.io/core/global-temp`) or you can find it from the book's GitHub repository (in the `chapter2` folder).

Then, we can load it up and pivot it:

```
temperatures = pd.read_csv('/Users/ben/Downloads/monthly_csv.csv')
temperatures['Date'] = pd.to_datetime(temperatures['Date'])
temperatures = temperatures.pivot('Date', 'Source', 'Mean')
```

Now we can use NumPy's polyfit functionality. It fits a polynomial of the form:

$$f(x) = \sum_{i=0}^{k} b_i x^{i+1}$$

In this formula, k is the degree of the polynomial and b is the coefficients we are trying to find.

It is just a function in NumPy to fit the coefficients. We can use the same function to fit seasonal variation and trend. Since trend can dominate over seasonality, before estimating seasonality, we remove the trend:

```
from numpy import polyfit

def fit(X, y, degree=3):
    coef = polyfit(X, y, degree)
    trendpoly = np.poly1d(coef)
    return trendpoly(X)

def get_season(s, yearly_periods=4, degree=3):
    X = [i%(365/4) for i in range(0, len(s))]
    seasonal = fit(X, s.values, degree)
    return pd.Series(data=seasonal, index=s.index)

def get_trend(s, degree=3):
    X = list(range(len(s)))
    trend = fit(X, s.values, degree)
    return pd.Series(data=trend, index=s.index)
```

Let's plot seasonality and trend on top of our global temperature increases!

```
import seaborn as sns

plt.figure(figsize=(12, 6))
```

```
temperatures['trend'] = get_trend(temperatures['GCAG'])
temperatures['season'] = get_season(temperatures['GCAG'] -
temperatures['trend'])
sns.lineplot(data=temperatures[['GCAG', 'season', 'trend']])
plt.ylabel('Temperature change');
```

This is the graph that we get:

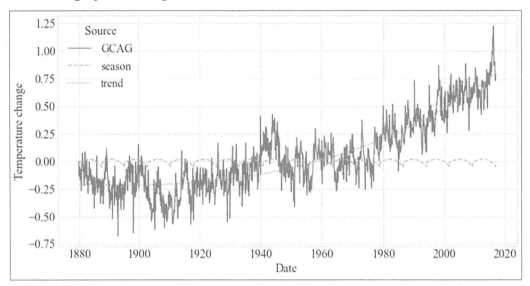

Figure 2.9: Temperature change from the late 19th century to today

This was to show that you can use plug-in functionality in NumPy for curve fitting in order to find both trend and seasonality. If you want to experiment further, you can play with the degree of the polynomial or with the seasonality component to see if you can get a better fit, or find another seasonality component. We could have used functionality from other libraries such as `seasonal.seasonal_decompose()` in `statsmodels`, or Facebook's Prophet, which decomposes using Fourier coefficients for the seasonal components.

Now that we've seen how to estimate seasonality and trend, let's move on to other statistics and visualizations. Continuing with the pollution dataset, and picking up the EEG dataset we saw in *Chapter 1*, we'll show practically in Python how to get these statistics and plots, and how to identify trend and seasonality.

Autocorrelation is the correlation of a signal with a lagged version of itself. The autocorrelation plot draws the autocorrelation as a function of lag. The autocorrelation plot can help find repeating patterns, and is often used in signal processing. The autocorrelation can help spot a periodic signal. Let's plot the autocorrelation of the pollution data:

```
pollution = pollution.pivot("Year", "City", "SPM")
pd.plotting.autocorrelation_plot(pollution['London'])
```

Here's the plot that we get:

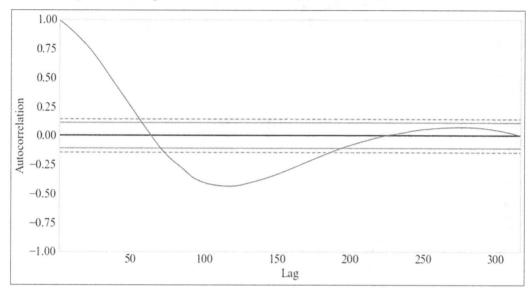

Figure 2.10: Autocorrelaton plot of pollution in London

We can see high autocorrelations with a lag of only a few years. There is a negative autocorrelation at around 100 years, after which point the autocorrelation stays around 0.

The plot of SPM clearly shows that air pollution is not a stationary process, since the autocorrelation is not flat. You can also compare the run of pollution that shows there's a **trend**, and therefore the mean also changes – another indication that the series is not stationary.

We can also test this statistically. A test for stationarity is the augmented Dickey–Fuller test:

```
from statsmodels.tsa import stattools

stattools.adfuller(pollution['London'])
```

```
(-0.33721640804242853,
 0.9200654843183897,
 13,
 303,
```

```
{'1%': -3.4521175397304784,
 '5%': -2.8711265007266666,
 '10%': -2.571877823851692},
1684.6992663493872)
```

The second return value is the p-value that gives the significance or the probability of obtaining test results at least as extreme as the observation given the null hypothesis. With p-values below 5% or 0.05 we would typically reject the null hypothesis, and we could assume that our time-series is stationary. In our case, we can't assume that the series is stationary.

We saw the graph of **electroencephalography** (**EEG**) signals in *Chapter 1, Introduction to Time-Series with Python,* and we mentioned that EEG signals show brain waves at several frequency ranges.

We can visualize this nicely. Let's go through it step by step in Python. We first need to do a few imports:

```python
import pandas as pd
import matplotlib.pyplot as plt
from matplotlib.dates import DateFormatter
import seaborn as sns
from sklearn.datasets import fetch_openml
```

OpenML is a project that provides benchmark datasets and a website for comparison of machine learning algorithms. The scikit-learn library provides an interface to OpenML that allows us to fetch data from OpenML. The whole measurement spans 117 seconds. So we need to set this up correctly as an index in pandas:

```python
eeg = fetch_openml(data_id=1471, as_frame=True)
increment = 117 / len(eeg['data'])
import numpy as np
index = np.linspace(
    start=0,
    stop=increment*len(eeg['data']),
    num=len(eeg['data'])
)
ts_index = pd.to_datetime(index, unit='s')
v1 = pd.Series(name='V1', data=eeg['data']['V1'].values, index=ts_
index)
```

We can slice our dataset directly. Please note that the `DatetimeIndex` is anchored in 1970, but we can ignore this safely here:

```
slicing = (v1.index >= '1970-01-01 00:00:08') & (v1.index <='1970-01-01
00:01:10.000000000')
v1[slicing]
```

Here's the slice:

```
1970-01-01 00:00:08.006208692     4289.74
1970-01-01 00:00:08.014019627     4284.10
1970-01-01 00:00:08.021830563     4280.00
1970-01-01 00:00:08.029641498     4289.74
1970-01-01 00:00:08.037452433     4298.46
                                    ...
1970-01-01 00:01:09.962547567     4289.74
1970-01-01 00:01:09.970358502     4283.08
1970-01-01 00:01:09.978169437     4284.62
1970-01-01 00:01:09.985980373     4289.23
1970-01-01 00:01:09.993791308     4290.77
Name: V1, Length: 7937, dtype: float64
```

This slicing avoids an artifact, a strong spike, occurring at around 1:20.

The graph we saw in *Chapter 1*, we can plot as follows:

```
date_formatter = DateFormatter("%S")
ax = v1[slicing].plot(figsize=(12, 6))
ax.xaxis.set_major_formatter(date_formatter)
plt.ylabel('voltage')
```

Here's the graph again:

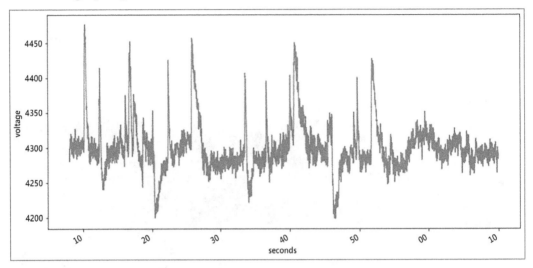

Figure 2.11: Voltage over time in an EEG signal

This is the plot of the EEG signal over time.

We can also resample the data to look at the series more coarsely, with less resolution, for example like this:

```
plt.subplot(311)
ax1 = v1[slicing].resample('1s').mean().plot(figsize=(12, 6))
ax1.xaxis.set_major_formatter(date_formatter)
plt.subplot(312)
ax1 = v1[slicing].resample('2s').mean().plot(figsize=(12, 6))
ax1.xaxis.set_major_formatter(date_formatter)
plt.subplot(313)
ax2 = v1[slicing].resample('5s').mean().plot(figsize=(12, 6))
ax2.xaxis.set_major_formatter(date_formatter)
plt.xlabel('seconds');
```

This is the graph with three subplots we get from resampling to frequencies of 1 second, 2 seconds, and 5 seconds, respectively:

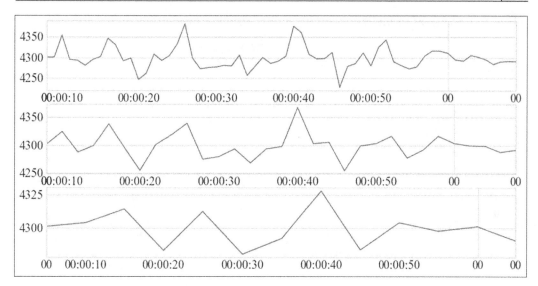

Figure 2.12: Resampled EEG signals

Each of the resampled signals in the plot could be more or less useful for analysis depending on the application. For high-frequency analysis, we shouldn't resample at all, while if we are trying to remove as much noise as possible, we should resample to a more coarse time resolution.

We can look at cyclic activity on a plot of spectral density. We can do this by applying a Fourier transform. Here, we go with the Welch method, which averages over time before applying the discrete Fourier transform:

```
from scipy import signal
fs = len(eeg['data']) // 117
f, Pxx_den = signal.welch(
    v1[slicing].values,
    fs,
    nperseg=2048,
    scaling='spectrum'
)
plt.semilogy(f, Pxx_den)
plt.xlabel('frequency [Hz]')
plt.ylabel('PSD [V**2/Hz]')
```

The spectral density plot, the **periodogram**, looks like this:

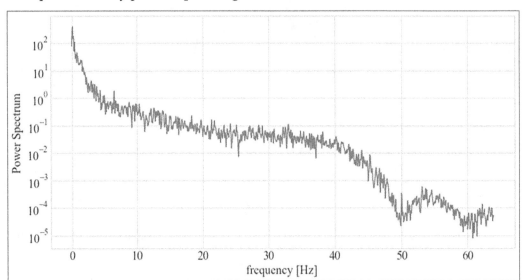

Figure 2.13: Periodogram of the EEG signals

The information in this plot is like the autocorrelation plot that we drew for pollution, however, it gives us information about how prominent certain frequencies are. In this case we see that low frequencies are particularly powerful. In other words, the signal shows a slow oscillation.

This brings the chapter to an end. Let's summarize what we've covered.

Summary

In this chapter, we introduced TSA as the process of extracting summary and other statistical information from time-series. We broke this process down into understanding the variables, uncovering relationships between variables, and identifying trend and seasonality.

We introduced datetime and pandas, the libraries *sine qua non* in TSA, and their functionalities for time-series; for example, resampling. Throughout the chapter, we listed and defined many summary statistics including mean, standard deviation, median, SE, confidence interval, Pearson correlation, and covariance.

We also talked about the concepts of seasonality, cyclic variation, and stationarity. We discussed why stationarity is important, and how to test for stationarity.

We also showed plotting functionality with Matplotlib and Seaborn, and how to generate different plots such as run charts, temporal line charts, correlation heatmaps, histograms, scatter plots, autocorrelation plots, and periodograms. In the practical example, we used an autocorrelation plot, which shows the correlation between different time steps, and the periodogram, which visualizes the power spectral density.

3

Preprocessing Time-Series

Preprocessing is a crucial step in machine learning that is nonetheless often neglected. Many books don't cover preprocessing in any depth or skip preprocessing entirely. When presenting to outsiders about a machine learning project, curiosity is naturally attracted to the algorithm rather than the dataset or the preprocessing.

One reason for the relative silence on preprocessing could be that it's less glamorous than machine learning itself. It is, however, often the step that takes the most time, sometimes estimated at around 98% of the whole machine learning process. And it is often in preprocessing that relatively easy work can have a great impact on the eventual performance of the machine learning model. The quality of the data goes a long way toward determining the outcome – low-quality input, in the worst case, can invalidate the machine learning work altogether (this is summarized in the adage "garbage in, garbage out").

Preprocessing includes curating and screening the data, something that overlaps with the analysis process covered in the previous chapter, *Chapter 2*, *Time-Series Analysis with Python*. The expected output of the preprocessing is a dataset on which it is easier to conduct machine learning. This can mean that it is more reliable and less noisy than the original dataset.

You can find the code for this chapter as a Jupyter notebook in the book's GitHub repository.

We're going to cover the following topics:

- What Is Preprocessing?
- Feature Transforms

- Feature Engineering
- Python Practice

We'll start off by discussing the basics of preprocessing.

What Is Preprocessing?

Anyone who's ever worked in a company on a machine learning project knows that real-world data is messy. It's often aggregated from multiple sources or using multiple platforms or recording devices, and it's incomplete and inconsistent. In preprocessing, we want to improve the data quality to successfully apply a machine learning model.

Data preprocessing includes the following set of techniques:

- Feature transforms
  - Scaling
  - Power/log transforms
  - Imputation
- Feature engineering

These techniques fall largely into two classes: either they tailor to the assumptions of the machine learning algorithm (feature transforms) or they are concerned with constructing more complex features from multiple underlying features (feature engineering). We'll only deal with univariate feature transforms, transforms that apply to one feature at a time. We won't discuss multivariate feature transforms (data reduction) such as variable selection or dimensionality reduction since they are not particular to time-series datasets.

Missing values are a common problem in machine learning, so we will discuss replacing missing values (imputation) in this chapter as well.

We'll be talking about features as the elementary units of preprocessing. We want to create input features for our machine learning process that make the model easier to train, easier to evaluate, or to improve the quality of model predictions. Our goal is to have features that are predictive of the target, and decorrelated (not redundant between themselves). Decorrelation is a requirement for linear models, but less important for more modern, for example tree-based, algorithms.

Although we'll mostly deal with feature engineering, we'll also mention target transformations. We could refer to target transformations more specifically as target engineering; however, since methods applied on targets are the same methods as those applied to features, I've included them under the same headings about feature engineering or feature transformations.

Please note that we define the main goal of our preprocessing as increasing the predictiveness of our features, or, in other words, we want to elevate the quality of our machine learning model predictions. We could have alternatively defined data quality in terms of accuracy, completeness, and consistency, which would have cast a much wider net including data aggregation and cleaning techniques, and methods of data quality assessment.

In this chapter, we are pragmatically reducing the scope of the treatment here to usefulness in machine learning. If our model is not fit for purpose, we might want to repeat or improve data collection, do more feature engineering, or build a better model. This again emphasizes the point that data analysis, preprocessing, and machine learning are an iterative process.

Binning or discretization can be a part of preprocessing as well, but can also be used for grouping data points by their similarity. We'll be discussing discretization together with other clustering techniques in *Chapter 6, Unsupervised Methods for Time-Series*.

Before we continue, let's go through some of the basics of preprocessing time-series datasets with Python. This will cover the theory behind operations with time-series data as an introduction.

Feature Transforms

Many models or training processes depend on the assumption that the data is distributed according to the normal distribution. Even the most widely used descriptors, the arithmetic mean and standard deviation, are largely useless if your dataset has a skew or several peaks (multi-modal). Unfortunately, observed data often doesn't fall within the normal distribution, so that traditional algorithms can yield invalid results.

When data is non-normal, transformations of data are applied to make the data as normal-like as possible and, thus, increase the validity of the associated statistical analyses.

Often it can be easier to eschew traditional machine learning algorithms of dealing with time-series data and, instead, use newer, so-called non-linear methods that are not dependent on the distribution of the data.

As a final remark, while all the following transformations and scaling methods can be applied to features directly, an interesting spin with time-series datasets is that they change over time, and we might not have full knowledge of the time-series. Many of these transformations have online variants, where all statistics are estimated and adjusted on the fly. You can take a look at *Chapter 8, Online Learning for Time-Series*, for more details on this topic.

In the next section, we'll look at scaling, which is a general issue in regression.

Scaling

Some features have natural bounds such as the age of a person or the year of production of a product. If these ranges differ between features, some model types (again, mostly linear models) struggle with this, preferring similar ranges, and similar means.

Two very common scaling methods are min-max scaling and z-score normalization.

Min-max scaling involves restricting the range of the feature within two constants, a and b. This is defined as follows:

$$x' = b \frac{a + x - \min(x)}{\max(x) - \min(x)}$$

In the special case that a is 0 and b is 1, this restricts the range of the feature within 0 and 1.

Z-score normalization is setting the mean of the feature to 0 and the variance to 1 (unit variance) like this:

$$x' = \frac{x - \bar{x}}{\sigma}$$

In the case that x comes from a Gaussian distribution, x' is a standard normal distribution.

In the next section, we'll look at log and power transformations. These transformations are quite important, especially for the traditional time-series models that we'll come across in *Chapter 5, Time-Series Forecasting with Moving Averages and Autoregressive Models*.

Log and Power Transformations

Both log and power transformations can compress values that spread over large magnitudes into a narrow range of output values. A **log transformation** is a feature transformation in which each value x gets replaced by $log(x)$.

The log function is the inverse of the exponential function, and it's important to remember that the range between 0 and 1 gets mapped to negative numbers $(-\infty, 0)$, while numbers $x>=1$ get compressed in the positive range. The choice of the logarithm is usually between the natural and base 10 but can be anything that can help so that your feature becomes closer to the symmetric bell-shaped distribution, which is the normal distribution.

Log transformation is, arguably, one of the most popular among the different types of transformations applied to take the distribution of the data closer to a Gaussian distribution. Log transformation can be used to reduce the skew of a distribution. In the best scenario, if the feature follows a log-normal distribution, then the log-transformed data follows a normal distribution. Unfortunately, your feature might not be distributed according to a log-normal distribution, so applying this transformation doesn't help.

Generally, I'd recommend exercising caution with data transformations. You should always inspect your data before and after the transformation. You want the variance of your feature to capture that of the target, so you should make sure you are not losing resolution. Further, you might want to check your data conforms – as should be the goal – more closely to the normal distribution. Many statistical methods have been developed to test the normality assumption of observed data, but even a simple histogram can give a great idea of the distribution.

Power transformations are often applied to transform the data from its original distribution to something that's more like a normal distribution. As discussed in the introduction, this can make a huge difference to the machine learning algorithm's ability to find a solution.

Power transforms are transformations preserving the original order (this property is called monotonicity) using power functions. A **power function** is a function of this form:

$$f(x) = cx^n$$

where $c \neq 0$.

When n is an integer and bigger than 1, we can make two major distinctions depending on whether n is odd or even. If it is even, the function x^n will tend toward positive infinity with large x, either positive or negative.

If it is odd, *f(x)* will tend toward positive infinity with increasing *x*, but toward negative infinity with *x*.

A power transformation is generally defined like this:

$$x_i^{(\lambda)} = \begin{cases} \dfrac{x_i^{\lambda} - 1}{\lambda(GM(x))^{\lambda-1}} & \text{if } \lambda \neq 0 \\ GM(x) \ln y_i & \text{if } \lambda = 0 \end{cases}$$

where *GM(x)* is the geometric mean of *x*:

$$GM(x) = \left(\prod_{i=1}^{n} x_i\right)^{\frac{1}{n}} = \sqrt[n]{x_1 x_2 \cdots x_n}$$

This reduces the transformation to the optimal choice of the parameter λ. For practical purposes, two power transformations are most commonly used:

- Box-Cox transformation
- Yeo–Johnson

For **Box-Cox transformation**, there are two variants, the one-parameter variant and the two-parameter variant. One-parameter Box–Cox transformations are defined like this:

$$x_i^{(\lambda)} = \begin{cases} \dfrac{x_i^{\lambda} - 1}{\lambda} & \text{if } \lambda \neq 0, \\ \ln x_i & \text{if } \lambda = 0, \end{cases}$$

The value of the parameter λ can be via different optimization methods such as the maximum likelihood that the transformed feature is Gaussian.

So the value of lambda corresponds to the exponent of the power operation, for example, X^{-3} with $\lambda = -3$ or $\sqrt{X}$ with $\lambda = 0.5$.

 Fun fact: Box-Cox transformation is named after statisticians George E.P Box and David Cox, who decided they had to work together because Box-Cox would sound good.

Yeo–Johnson transformation is an extension of Box-Cox transformation that allows zero and negative values of *x*. λ can be any real number, where $\lambda = 1$ produces the identity transformation. The transformation is defined as:

$$x_i^{(\lambda)} = \begin{cases} \dfrac{(x_i + 1)^\lambda - 1}{\lambda} & \text{if } \lambda \neq 0, x \geq 0 \\[2ex] \log(x_i + 1) & \text{if } \lambda = 0, x \geq 0 \\[2ex] -[\dfrac{(-x_i + 1)^{(2-\lambda)} - 1]}{2 - \lambda}) & \text{if } \lambda \neq 2, x < 0 \\[2ex] -\log(-x_i + 1) & \text{if } \lambda = 2, x < 0 \end{cases}$$

Finally, **quantile transformation** can map a feature to the uniform distribution based on an estimate of the cumulative distribution function. Optionally, this can then be mapped in a second step to normal distribution. The advantage of this transform, similar to other transformations that we've talked about, is that it makes features more convenient to process and plot, and easier to compare, even if they were measured at different scales.

In the next section, we'll look at imputation, which literally means the assignment of values by inference, but in machine learning, often is more narrowly meant to refer to replacing missing values.

Imputation

Imputation is the replacement of missing values. This is important for any machine learning algorithm that can't handle missing values. Generally, we can distinguish the following types of imputation techniques:

- Unit imputation – where missing values are replaced by a constant such as the mean or 0
- Model-based imputation – where missing values are replaced with predictions from a machine learning model

Unit imputation is by far the most popular imputation technique, partly because it's very easy to do, and because it's less heavy on computational resources than model-based imputation.

We'll do imputation in the practice section of this chapter. In the next section, we'll talk about feature engineering.

Feature Engineering

Machine learning algorithms can use different representations of the input features. As we've mentioned in the introduction, the goal of feature engineering is to produce new features that can help us in the machine learning process. Some representations or augmentations of features can boost performance.

We can distinguish between hand-crafted and automated feature extraction, where hand-crafted means that we look through the data and try to come up with representations that could be useful, or we can use a set of features that have been established from the work of researchers and practitioners before. An example of a set of established features is **Catch22**, which includes 22 features and simple summary statistics extracted from phase-dependant intervals. The Catch22 set is a subset of the **Highly Comparative Time-Series Analysis (HCTSA)** toolbox, another set of features.

Another distinction is between interpretable and non-interpretable features. Interpretable features could be summary features such as the mean, max, min, and others. These could be pooled within time periods, windows, to give us more features.

In features for time-series, a few preprocessing methods come with their recommended machine learning model. For example, a ROCKET model is a linear model on top of the ROCKET features.

Taken to the extreme, this can be a form of **model stacking**, where the outcomes of models serve as the inputs to other models. This can be an effective way of decomposing the learning problem by training less complex (fewer features, fewer parameters) models in a supervised setting and using their outputs for training other models.

Please note it is important that any new feature depends only on past and present inputs. In signal processing, this kind of operation is called a **causal filter**. The word causal indicates that the filter output for a value at time t only uses information available at time t and doesn't peek into the future. Conversely, a filter whose output also depends on future inputs is non-causal. We'll discuss Temporal Convolutional Networks, basically causal convolutions, in Chapter 10, Deep Learning for Time-Series.

We should take great care in training and testing that any statistics extracted and applied in preprocessing are carefully considered – at best, the model performance will be overly optimistic if it relies on data that shouldn't be available during prediction. We'll discuss leakage in the next chapter, *Chapter 4, Machine Learning for Time-Series*.

If we have many features, we might want to simplify our model building process by pruning the available features and using only a subset (feature selection), or instead, using a new set of features that describe the essential quality of the features (dimensionality reduction).

We can distinguish the following types of features with time-series:

- Date- and time-related features
    - Calendar features (date-related)
    - Time-related features
- Window-based features

Calendar- and time-related features are very similar, so we'll discuss them in the same section.

Window-based features are features that integrate features within a (rolling) window, that is, within a time period. Examples of these are averages over 15-minute windows or sales within 7 days. Since we dealt with rolling windows in *Chapter 2, Time-Series Analysis with Python*, in this chapter, we'll deal with more complex features such as convolutions and shapelets.

Many preprocessing algorithms are implemented in sktime. Another handy library is tsfresh, which calculates an enormous number of interpretable features for time-series. In the code in this chapter, we've accessed tsfresh features through feature tools.

Let's do some more time-series preprocessing in Python! We'll discuss date- and time-related features next.

Date- and Time-Related Features

Date and time variables contain information about dates, time, or a combination (datetime). We saw several examples in the previous chapter, *Chapter 2, Time-Series Analysis with Python* – one of them was the year corresponding to pollution. Other examples could be the birth year of a person or the date of a loan being taken out.

If we want to feed these fields into a machine learning model, we need to derive relevant information. We could feed the year as an integer, for example, but there are many more examples of extracted features from datetime variables, which we'll deal with in this section. We can significantly improve the performance of our machine learning model by enriching our dataset with these extracted features.

Workalendar is a Python module that provides classes able to handle calendars, including lists of bank and religious holidays, and it offers working-day-related functions. Python-holidays is a similar library, but here we'll go with workalendar.

In the next section, we'll discuss ROCKET features.

ROCKET

The research paper *ROCKET: Exceptionally fast and accurate time-series classification using random convolutional kernels* (Angus Dempster, François Petitjean, Geoffrey I. Webb; 2019) presents a novel methodology for convolving time-series data with random kernels that can result in higher accuracy and faster training times for machine learning models. What makes this paper unique is banking on the recent successes of convolutional neural networks and transferring them to preprocessing for time-series datasets.

We will go into more details of this paper. **ROCKET**, short for **RandOm Convolutional KErnel Transform**, is based on convolutions, so let's start with convolutions.

Convolutions are a very important transformation, especially in image processing, and are one of the most important building blocks of deep neural networks in image recognition. Convolutions consist of feedforward connections, called **filters** or **kernels**, that are applied to rectangular patches of the image (the previous layer). Each resulting image is then the sliding window of the kernel over the whole image. Simply put, in the case of images, a kernel is a matrix used to modify the images.

A sharpening kernel can look like this:

$$\begin{bmatrix} 0 & -1 & 0 \\ -1 & 5 & -1 \\ 0 & -1 & 0 \end{bmatrix}$$

If we multiply this kernel to all local neighborhoods in turn, we get a sharper image as illustrated below (original on the left, sharpened on the right):

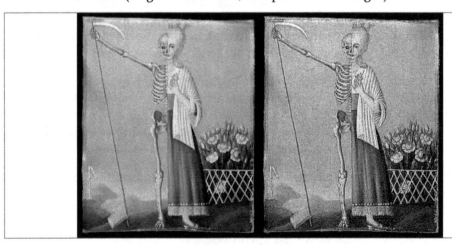

Figure 3.1: Sharpening filter

This picture is a gray version of "A woman divided into two, representing life and death" (owned by the Wellcome Collection, a museum and exhibition center in London; licensed under CC BY 4.0).

The sharpening kernel emphasizes differences in adjacent pixel values. You can see that the picture on the right is much grainier or vivid – a result of the convolution.

We can apply kernels not only to images but also to vectors or matrices, and this brings us back to ROCKET. ROCKET computes two aggregate features from each kernel and feature convolution. The two features are created using the well-known methodology global/average max pooling and a novel methodology that we'll come to in a second.

Global max pooling outputs the maximum value from the result of convolution and max pooling takes the maximum value within a pool size. For example, if the result of the convolution is 0,1,2,2,5,1,2, global max pooling returns 5, whereas **max pooling** with pool size 3 outputs the maxima within windows of 3, so we'll get 2,2,5,5,5.

Positive Proportion Value (PPV), the methodology from the paper, is the proportion (percentage) of values from the convolution that are positive (or above a bias threshold).

We can improve machine learning accuracy from time-series by applying transformations with convolutional kernels. Each feature gets transformed by random kernels, the number of which is a parameter to the algorithm. This is set to 10,000 by default. The transformed features can now be fed as input into any machine learning algorithm. The authors propose to use linear algorithms like ridge regression classifier or logistic regression.

The idea of ROCKET is very similar to Convolutional Neural Networks (CNNs), which we'll discuss in chapter 10, Deep Learning for Time-Series, however, two big differences are:

1. ROCKET doesn't use any hidden layers or non-linearities
2. The convolutions are applied independently for each feature

In the next section, we'll be discussing shapelets.

Shapelets

Shapelets for time-series were presented in the research paper *Time-Series Shapelets: A New Primitive for Data Mining* (Lexiang Ye and Eamonn Keogh, 2009). The basic idea of shapelets is decomposing time-series into discriminative subsections (shapelets).

In the first step, the shapelets are learned. The algorithm calculates the information gain of possible candidates and picks the best candidates to create a shapelet dictionary of discriminating subsections. This can be quite expensive. Then, based on the shapelet decomposition of the features, a decision tree or other machine learning algorithms can be applied.

Shapelets have several advantages over other methods:

- They can provide interpretable results
- The application of shapelets can be very fast – only depending on the matching of features against the dictionary of shapelets
- The performance of machine learning algorithms on top of shapelets is usually very competitive

It's time that we go through Python exercises with actual datasets.

Python Practice

NumPy and SciPy offer most of the functionality that we need, but we might need a few more libraries.

In this section, we'll use several libraries, which we can quickly install from the terminal, **the Jupyter Notebook**, or similarly **from Anaconda Navigator**:

```
pip install -U tsfresh workalendar astral "featuretools[tsfresh]"
sktime
```

All of these libraries are quite powerful and each of them deserves more than the space we can give to it in this chapter.

Let's start with log and power transformations.

Log and Power Transformations in Practice

Let's create a distribution that's not normal, and let's log-transform it. We'll plot the original and transformed distribution for comparison, and we'll apply a statistical test for normality.

Let's first create the distribution:

```
from scipy.optimize import minimize
import numpy as np
np.random.seed(0)
```

```
pts = 10000
vals = np.random.lognormal(0, 1.0, pts)
```

Values are sampled from a lognormal distribution. I've added a call to the random number generator seed function to make sure the result is reproducible for readers.

We can visualize our array as a histogram:

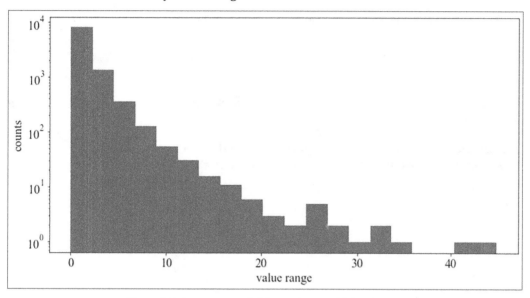

Figure 3.2: An array sampled from a lognormal distribution

I've used a log scale on the y-axis. We can see that the values spread over a number of orders of magnitude.

We can apply the standard normalization to z-scores. We can also apply a statistical normality test on one of the transformed distributions:

```
from sklearn.preprocessing import StandardScaler
from scipy.stats import normaltest
scaler = StandardScaler()
vals_ss = scaler.fit_transform(vals.reshape(-1, 1))
_, p = normaltest(vals_ss)
print(f"significance: {p:.2f}")
```

The null hypothesis of this statistical test is that the sample comes from a normal distribution. Therefore significance values (p-values) lower than a threshold, typically set to 0.05 or 0.01, would let us reject the null hypothesis.

We are getting this output: `significance: 0.00`.

We can conclude from the test that we are not getting a null distribution from our transformation by standard scaling.

This should be obvious, but let's get it out of the way: we are getting the same significance for the minmax transformed values:

```
from sklearn.preprocessing import minmax_scale
vals_mm = minmax_scale(vals)
_, p = normaltest(vals_mm.squeeze())
print(f"significance: {p:.2f}")
```

We therefore reach the same conclusion: the minmax transformation hasn't helped us get a normal-like distribution.

We can plot the original and the standard scaled distribution against each other. Unsurprisingly, visually, the two distributions look the same except for the scale.

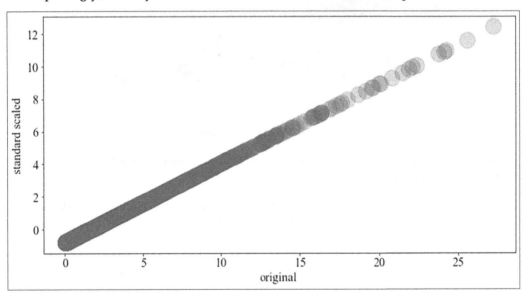

Figure 3.3: The linear transformation against the original values

We can see everything lies on the diagonal.

Let's use a log transformation:

```
log_transformed = np.log(vals)
_, p = normaltest(log_transformed)
print(f"significance: {p:.2f}")
```

We are getting a significance of `0.31`. This lets us conclude that we can't reject the null hypothesis. Our distribution is similar to normal. In fact, we get a standard deviation close to 1.0 and a mean close to 0.0 as we would expect with a normal distribution.

We can see that the log-normal distribution is a continuous probability distribution whose logarithm is normally distributed, so it's not entirely surprising that we get this result.

We can plot the histogram of the log-transformed distribution:

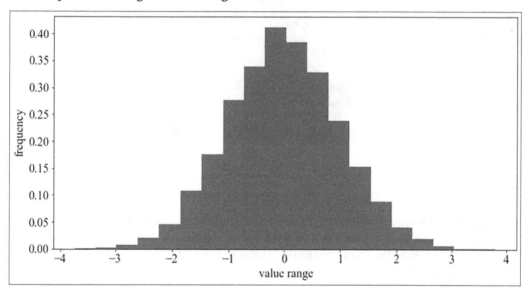

Figure 3.4: Log-transformed lognormal array

The log transform looks much more normal-like as we can appreciate.

We can also apply Box-Cox transformation:

```
from scipy.stats import boxcox
vals_bc = boxcox(vals, 0.0)
```

We are getting a significance of 0.46. Again, we can conclude that our Box-Cox transform is normal-like. We can also see this in a plot of the Box-Cox transformed distribution:

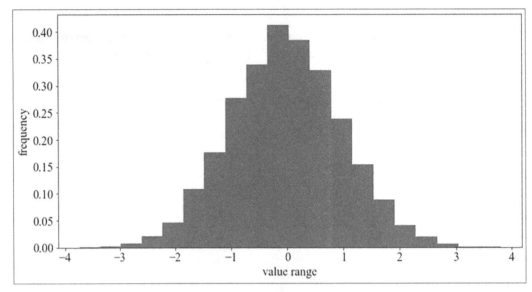

Figure 3.5: Box-Cox transformed lognormal array

Again, this looks very much like a normal distribution. This plot looks pretty much the same as the previous one of the log transformation, which shouldn't be surprising given that the log operation corresponds to a lambda parameter of 0 in the Box-Cox transformation.

This is a small selection of transformations that can help us reconcile our data with the common normality assumption in classical forecasting methods.

Let's look at imputation in practice.

Imputation

It is rather uncommon for machine learning algorithms to be able to deal with missing values directly. Rather, we'll either have to replace missing values with constants or infer probable values given the other features.

The scikit-learn documentation lists a simple example of unit imputation:

```
import numpy as np
from sklearn.impute import SimpleImputer
imp_mean = SimpleImputer(missing_values=np.nan, strategy='mean')
```

```
imp_mean.fit([[7, 2, 3], [4, np.nan, 6], [10, 5, 9]])
SimpleImputer()
df = [[np.nan, 2, 3], [4, np.nan, 6], [10, np.nan, 9]]
print(imp_mean.transform(df))
```

We are again using (similar to the standard scaler before) the scikit-learn transformers, which come with `fit()` and `transform()` methods.

We get the following imputed values:

```
[[ 7.   2.   3. ]
 [ 4.   3.5  6. ]
 [10.   3.5  9. ]]
```

The missing values are replaced with the mean of the columns.

Let's look at annotating holidays as derived date features.

Holiday Features

If we want to get the holidays for the United Kingdom, we can do this:

```
from workalendar.europe.united_kingdom import UnitedKingdom
UnitedKingdom().holidays()
```

We get the following holidays:

```
[(datetime.date(2021, 1, 1), 'New year'),
 (datetime.date(2021, 4, 2), 'Good Friday'),
 (datetime.date(2021, 4, 4), 'Easter Sunday'),
 (datetime.date(2021, 4, 5), 'Easter Monday'),
 (datetime.date(2021, 5, 3), 'Early May Bank Holiday'),
 (datetime.date(2021, 5, 31), 'Spring Bank Holiday'),
 (datetime.date(2021, 8, 30), 'Late Summer Bank Holiday'),
 (datetime.date(2021, 12, 25), 'Christmas Day'),
 (datetime.date(2021, 12, 26), 'Boxing Day'),
 (datetime.date(2021, 12, 27), 'Christmas Shift'),
 (datetime.date(2021, 12, 28), 'Boxing Day Shift')]
```

We can generate holidays by year and then look up holidays by date.

Similarly, we can get holidays for other places, for example, California, USA. We can also extract lists of holidays, and add custom holidays:

```python
from typing import List
from dateutil.relativedelta import relativedelta, TH
import datetime
from workalendar.usa import California
def create_custom_holidays(year: int) -> List:
    custom_holidays = California().holidays()
    custom_holidays.append((
        (datetime.datetime(year, 11, 1) + relativedelta(weekday=TH(+4))
+ datetime.timedelta(days=1)).date(),
        "Black Friday"
    ))
    return {k: v for (k, v) in custom_holidays}

custom_holidays = create_custom_holidays(2021)
```

This gives us our custom holidays for the year 2021:

```
{datetime.date(2021, 1, 1): 'New year',
 datetime.date(2021, 1, 18): 'Birthday of Martin Luther King, Jr.',
 datetime.date(2021, 2, 15): "Washington's Birthday",
 datetime.date(2021, 3, 31): 'Cesar Chavez Day',
 datetime.date(2021, 5, 31): 'Memorial Day',
 datetime.date(2021, 7, 4): 'Independence Day',
 datetime.date(2021, 7, 5): 'Independence Day (Observed)',
 datetime.date(2021, 9, 6): 'Labor Day',
 datetime.date(2021, 11, 11): 'Veterans Day',
 datetime.date(2021, 11, 25): 'Thanksgiving Day',
 datetime.date(2021, 11, 26): 'Thanksgiving Friday',
 datetime.date(2021, 12, 24): 'Christmas Day (Observed)',
 datetime.date(2021, 12, 25): 'Christmas Day',
 datetime.date(2021, 12, 31): 'New Years Day (Observed)',
 datetime.date(2016, 11, 25): 'Black Friday'}
```

Please note that we are using type hints in the code segment above. We are declaring the signature of our function like this:

```python
def create_custom_holidays(year: int) -> List:
```

This means we are expecting an integer called year and we are expecting a List as an output. Annotations are optional and they are not getting checked (if you don't invoke mypy), but they can make code in Python much clearer.

Now we can implement a simple lookup like this:

```
def is_holiday(current_date: datetime.date):
    """Determine if we have a holiday."""
    return custom_holidays.get(current_date, False)

today = datetime.date(2021, 4, 11)
is_holiday(today)
```

I am getting a False even though I wish it were a holiday.

This can be a very useful feature for a machine learning model. For example, we could imagine a different profile of users who apply for loans on bank holidays or on a weekday.

Date Annotation

The calendar module offers lots of methods, for example, monthrange() - calendar. monthrange returns the first weekday of the month and the number of days in a month for a given year and month. The day of the week is given as an integer, where Monday is 0 and Sunday is 6.

```
import calendar
calendar.monthrange(2021, 1)
```

We should be getting (4, 31). This means the first weekday of 2021 was a Friday. January 2021 had 31 days.

We can also extract features relevant to the day with respect to the year. The following function provides the number of days since the end of the previous year and to the end of the current year:

```
from datetime import date
def year_anchor(current_date: datetime.date):
    return (
        (current_date - date(current_date.year, 1, 1)).days,
        (date(current_date.year, 12, 31) - current_date).days,
    )

year_anchor(today)
```

This feature could provide a general idea of how far into the year we are. This can be useful both for estimating a trend and for capturing cyclic variations.

Similarly, we can extract the number of days from the first of the month and to the end of the month:

```
def month_anchor(current_date: datetime.date):
    last_day = calendar.monthrange(current_date.year, current_date.
month)[0]

    return (
        (current_date - datetime.date(current_date.year, current_date.
month, 1)).days,
        (current_date - datetime.date(current_date.year, current_date.
month, last_day)).days,
    )

month_anchor(today)
```

A feature like this could also provide some useful information. I am getting (10, 8).

In retail, it is very important to predict the spending behavior of customers. Therefore, in the next section, we'll write annotation for pay days.

Paydays

We could imagine that some people get paid in the middle or at the end of the month, and would then access our website to buy our products.

Most people would get paid on the last Friday of the month, so let's write a function for this:

```
def get_last_friday(current_date: datetime.date, weekday=calendar.
FRIDAY):
    return max(week[weekday]
        for week in calendar.monthcalendar(
            current_date.year, current_date.month
        ))

get_last_friday(today)
```

I am getting 30 as the last Friday.

Seasons can also be predictive.

Seasons

We can get the season for a specific date:

```
YEAR = 2021
seasons = [
    ('winter', (date(YEAR,  1,  1),  date(YEAR,  3, 20))),
    ('spring', (date(YEAR,  3, 21),  date(YEAR,  6, 20))),
    ('summer', (date(YEAR,  6, 21),  date(YEAR,  9, 22))),
    ('autumn', (date(YEAR,  9, 23),  date(YEAR, 12, 20))),
    ('winter', (date(YEAR, 12, 21),  date(YEAR, 12, 31)))
]

def is_in_interval(current_date: datetime.date, seasons):
    return next(season for season, (start, end) in seasons
            if start <= current_date.replace(year=YEAR) <= end)

is_in_interval(today, seasons)
```

We should be getting spring here, but the reader is encouraged to try this with different values.

The Sun and Moon

The Astral module offers information about sunrise, moon phases, and more. Let's get the hours of sunlight for a given day in London:

```
from astral.sun import sun
from astral import LocationInfo
CITY = LocationInfo("London", "England", "Europe/London", 51.5, -0.116)
def get_sunrise_dusk(current_date: datetime.date, city_name='London'):
    s = sun(CITY.observer, date=current_date)
    sunrise = s['sunrise']
    dusk = s['dusk']
    return (sunrise - dusk).seconds / 3600

get_sunrise_dusk(today)
```

I am getting 9.788055555555555 hours of daylight.

It can often be observed that the more hours of daylight, the more business activity. We could speculate that this feature could be helpful in predicting the volume of our sales.

Business Days

Similarly, if a month has more business days, we could expect more sales for our retail store. On the other hand, if we are selling windsurfing lessons, we might want to know the number of holidays in a given month. The following function extracts the number of business days and weekends/holidays in a month:

```
def get_business_days(current_date: datetime.date):
    last_day = calendar.monthrange(current_date.year, current_date.
month)[1]
    rng = pd.date_range(current_date.replace(day=1), periods=last_
day, freq='D')
    business_days = pd.bdate_range(rng[0], rng[-1])
    return len(business_days), last_day - len(business_days)


get_business_days(date.today())
```

We should be getting (22, 9) – 22 business days and 9 weekend days and holidays.

Automated Feature Extraction

We can also use automated feature extraction tools from modules like featuretools. Featuretools calculates many datetime-related functions. Here's a quick example:

```
import featuretools as ft
from featuretools.primitives import Minute, Hour, Day, Month, Year,
Weekday

data = pd.DataFrame(
    {'Time': ['2014-01-01 01:41:50',
              '2014-01-01 02:06:50',
              '2014-01-01 02:31:50',
              '2014-01-01 02:56:50',
              '2014-01-01 03:21:50'],
     'Target': [0, 0, 0, 0, 1]}
)
data['index'] = data.index
es = ft.EntitySet('My EntitySet')
es.entity_from_dataframe(
```

```
        entity_id='main_data_table',
        index='index',
        dataframe=data,
        time_index='Time'
    )
fm, features = ft.dfs(
        entityset=es,
        target_entity='main_data_table',
        trans_primitives=[Minute, Hour, Day, Month, Year, Weekday]
    )
```

Our features are Minute, Hour, Day, Month, Year, and Weekday. Here's our DataFrame, fm:

index	Target	DAY(Time)	HOUR(Time)	MINUTE(Time)	MONTH(Time)	WEEKDAY(Time)	YEAR(Time)
0	0	1	1	41	1	2	2014
1	0	1	2	6	1	2	2014
2	0	1	2	31	1	2	2014
3	0	1	2	56	1	2	2014
4	1	1	3	21	1	2	2014

Figure 3.6: Featuretools output

We could extract many more features. Please see the featuretools documentation for more details.

The tsfresh module also provides automated functionality for feature extraction:

```
from tsfresh.feature_extraction import extract_features
from tsfresh.feature_extraction import ComprehensiveFCParameters

settings = ComprehensiveFCParameters()
extract_features(data, column_id='Time', default_fc_parameters=settings)
```

 Please note that tsfresh optimizes features using statsmodels' autoregression, and (last I checked) still hasn't been updated to use statsmodels.tsa.AutoReg instead of statsmodels.tsa.AR, which has been deprecated.

We get 1,574 features that describe our time object. These features could help us in machine learning models.

Let's demonstrate how to extract ROCKET features from a time-series.

ROCKET

We'll be using the implementation of ROCKET in the `sktime` library.

The `sktime` library represents data in a nested DataFrame. Each column stands for a feature, as expected, however, what may be surprising is that each row is an instance of a time-series. Each cell contains an array of all values for a given feature over time. In other words, each cell has a nested object structure, where instance-feature combinations are stored.

This structure makes sense, because it allows us to store multiple instances of time-series in the same DataFrame, however, it's not intuitive at first. Fortunately, SkTime provides utility functions to unnest the SkTime datasets, as we will see.

If we want to load an example time-series in SkTime, we can do this:

```
from sktime.datasets import load_arrow_head
from sktime.utils.data_processing import from_nested_to_2d_array
X_train, y_train = load_arrow_head(split="train", return_X_y=True)
from_nested_to_2d_array(X_train).head()
```

We get an unnested DataFrame like this:

	dim_0__0	dim_0__1	dim_0__2	dim_0__3	dim_0__4	dim_0__5	dim_0__6	dim_0__7	dim_0__8	dim_0__9	...	d
0	-1.9630	-1.9578	-1.9561	-1.9383	-1.8967	-1.8699	-1.8387	-1.8123	-1.7364	-1.6733	...	
1	-1.7746	-1.7740	-1.7766	-1.7307	-1.6963	-1.6574	-1.6362	-1.6098	-1.5434	-1.4862	...	
2	-1.8660	-1.8420	-1.8350	-1.8119	-1.7644	-1.7077	-1.6483	-1.5826	-1.5315	-1.4936	...	
3	-2.0738	-2.0733	-2.0446	-2.0383	-1.9590	-1.8745	-1.8056	-1.7310	-1.7127	-1.6280	...	
4	-1.7463	-1.7413	-1.7227	-1.6986	-1.6772	-1.6304	-1.5794	-1.5512	-1.4740	-1.4594	...	

5 rows × 251 columns

Figure 3.7: ROCKET features

Again, each row is a time-series. There's only one feature, called `dim_0`. The time-series has 251 measurements.

We can import ROCKET, and then create the ROCKET features. We'll first have to learn the features and then apply them. This is a typical pattern for machine learning. In scitkit-learn, we'd use the `fit()` and `predict()` methods for models, where `fit()` is applied on the training data and `predict()` gives the predictions on a test set.

The learning step should only ever be applied to the training set. One of the parameters in ROCKET is the number of kernels. We'll set it to 1,000 here, but we can set it to a higher number as well. 10,000 kernels is the default:

```
from sktime.transformations.panel.rocket import Rocket
rocket = Rocket(num_kernels=1000)
rocket.fit(X_train)
X_train_transform = rocket.transform(X_train)
```

The returned dataset is not nested, and it contains 2,000 columns. Each column describes the whole time-series but is the result from a different kernel.

In the next section, we'll do a shapelets exercise.

Shapelets in Practice

Let's create shapelets for the dataset we used before when we looked at ROCKET. We'll again use SkTime:

```
from sktime.transformations.panel.shapelets import
ContractedShapeletTransform
shapelets_transform = ContractedShapeletTransform(
    time_contract_in_mins=1,
    num_candidates_to_sample_per_case=10,
    verbose=0,
)
shapelets_transform.fit(X_train, y_train)
```

The training could take a few minutes. We'll get some output about the candidates that are being examined.

We can again transform our time-series using our shapelet transformer. This works as follows:

```
X_train_transform = shapelets_transform.transform(X_train)
```

This gives us a transformed dataset that we can use in machine learning models. We encourage the reader to play around with these feature sets.

This concludes our Python practice.

Summary

Preprocessing is a crucial step in machine learning that is often neglected. Many books don't cover preprocessing as a topic or skip preprocessing entirely. However, it is often in preprocessing that relatively easy wins can be achieved. The quality of the data determines the outcome.

Preprocessing includes curating and screening the data. The expected output of the preprocessing is a dataset on which it is easier to conduct machine learning. This can mean that it is more reliable and less noisy than the original dataset.

We've talked about feature transforms and feature engineering approaches to time-series data, and we've talked about automated approaches as well.

In the next chapters, we'll explore how we can use these extracted features in a machine learning model. We'll discuss combinations of features and modeling algorithms in the next chapter, *Chapter 4, Introduction to Machine Learning for Time-Series*. In *Chapter 5, Time-Series Forecasting with Moving Averages and Autoregressive Models*, we'll be using machine learning pipelines, where we can connect feature extraction and machine learning models.

4

Introduction to Machine
Learning for Time-Series

In previous chapters, we've talked about time-series, time-series analysis, and preprocessing. In this chapter, we'll talk about machine learning for time-series. **Machine learning** is the study of algorithms that improve through experience. These algorithms or models can make systematic, repeatable, validated decisions based on data. This chapter is meant to give an introduction given both the context and the technical background to much of what we'll use in the remainder of this book.

We'll go through different kinds of problems and applications of machine learning in time-series, and types of analyses relevant to machine learning and time-series analysis. We'll explain the main machine learning problems with time-series, such as forecasting, classification, regression, segmentation, and anomaly detection. We'll then review the basics of machine learning as relevant to time-series. Then, we'll look at the history and current uses of machine learning for time-series.

We're going to cover the following topics:

- Machine learning with time-series
    - Supervised, unsupervised, and reinforcement learning
    - History of Machine Learning
- Machine learning workflow
    - Cross-validation
    - Error metrics for time-series
    - Comparing time-series

- Machine learning algorithms for time-series

We'll start with a general introduction to machine learning with time-series.

Machine learning with time-series

In this section, I'll give an introduction to applications and the main categories of machine learning with time-series.

Machine learning approaches for time-series are crucial in domains such as economics, medicine, meteorology, demography, and many others. Time-Series datasets are ubiquitous and occur in domains as diverse as healthcare, economics, social sciences, Internet-of-Things applications, operations management, digital marketing, cloud infrastructure, the simulation of robotic systems, and others. These datasets are of immense practical importance, as they can be leveraged to forecast and predict the detection of anomalies more effectively, thereby supporting decision making.

The technical applications within machine learning for time-series abound in techniques. A few applications are as follows:

- Curve fitting
- Regression
- Classification
- Forecasting
- Segmentation/clustering
- Anomaly detection
- Reinforcement learning

We will examine these technical applications in this book. These different applications have different statistical methods and models behind them that can overlap.

Let's go briefly through some of these applications for an overview of what to expect in the chapters to come.

Curve fitting is the task of fitting a mathematical function (a curve) to a series of points. The mathematical function is defined by parameters, and the parameters are adapted to fit the time-series through optimization. Curve fitting can be employed as a visual aid on graphs or for inference (extrapolation).

Regression is an umbrella term for statistical approaches for finding relationships between independent variables (features) and independent variables (targets). For instance, we could be predicting the exact temperature based on the release of carbon dioxide and methane. If there's more than one outcome variable, this is called multi-target (or multi-output). An example of this could be predicting the temperature for different locations at the same time.

When the problem is assigning labels to a time-series (or a part of it), this is called **classification**. The main difference to regression is that the prediction is categorical rather than continuous. The model that's used for classification is often referred to as a classifier. Classification can be binary, when there are precisely two classes, or multi-class, when there are more categories. An example would be detecting eye movements or epilepsy in EEG signals.

Making predictions about the future is called **forecasting**. Forecasting can be based only on the time-series values itself or on other variables. The techniques can range from curve fitting to extrapolating, from analysis of the current trends and variability to complex machine learning techniques. For example, we could be forecasting global temperatures based on the data of the last 100 years, or we could be forecasting the economic wellbeing of a nation. The antonym of forecasting is **backcasting**, where we make predictions about the past. We could be backcasting temperatures backward in time from before we have data available.

Segmentation, or **clustering**, is the process of grouping parts of the time-series into clusters (or segments) of different regimes, behaviors, or baselines. An example would be different activity levels in brain waves.

Within the context of time-series, **anomaly detection**, also known as **outlier detection**, is the task of identifying events that are rare or outside the norm. These could be novel, changes of regime, noise, or just exceptions. A rather crude example would be an outage of the electricity grid detectable in a sudden drop of the voltage. More subtle perhaps, by way of an example, would be an increase in the number of calls to a call center within a certain period. In both cases, anomaly detection could provide actionable business insights. Techniques in anomaly detection can range from simple thresholding or statistics to a set of rules, to pattern-based approaches based on the time-series distribution.

Finally, **reinforcement learning** is the practice of learning based on maximizing expected rewards from a series of decisions. Reinforcement learning algorithms are employed in environments where there's a high level of uncertainty. This could mean that the conditions are unstable (high variation) or there's a general lack of information. Applications are bidding and pricing algorithms in stock trading or general auctions, and control tasks.

Let's dive into a bit more detail about what these terms mean.

Supervised, unsupervised, and reinforcement learning

Machine learning can be broadly categorized into supervised, unsupervised, and reinforcement learning, as this diagram shows:

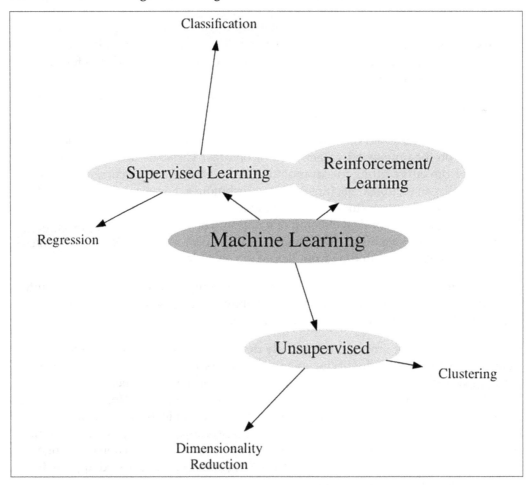

Figure 4.1: Dividing machine learning into categories

In **supervised** learning, the features are mapped to outcomes $y \in Y$ in a process called **prediction** (sometimes **inference**).

In the supervised case, parameters are estimated from labeled observations. We need to have the outcome available for each observation as the **target** column (or columns, in the plural).

Therefore, the machine learning algorithm finds a mapping from X to Y.

$$f: X \rightarrow Y$$

The function $f \in F$ is just one possible mapping or model of the input distribution X to the output distribution Y.

Supervised machine learning can be categorized into classification and regression. In regression, our targets are continuous, and the goal is to predict the value.

$$Y \subset \mathbb{R}^{\{N,\cdot\}}$$

The target Y can be real-valued, either a single value or a higher dimensionality (multioutput).

The labels match the dataset in length, but there can be several labels for each observation as well (multi-output).

An example would be the number of products sold in a shop on a specific day or the amount of oil coming through a pipeline over the next month. The features could include current sales, demand, or day of the week.

In **classification**, the goal is to predict the class of the observation. In this case, $y \in Y$ could be drawn from a categorical distribution, a distribution consisting of ordinal values, for example, integers as in $k \in \{1 \dots K\}$.

Sometimes, we want to find a function that would give us probabilities or scores for given observations:

$$f: X \times Y \rightarrow \mathbb{R}$$

In practical terms, regression and classification are very similar problems, and often regression and classification algorithms can be applied interchangeably. However, understanding the distinction is crucial to dealing with any specific problem in an appropriate manner.

In time-series **forecasting**, the historical values are extrapolated into the future. The only features are the past values. For example, we could be estimating the calls into a call center over the next month based on calls during the last 2 years. The forecasting task could be univariate, relying on and extrapolating a single feature, or multivariate, where multiple features are projected into the future.

In **unsupervised** learning, the algorithm's task is to categorize observations based on their features. Examples of unsupervised learning are clustering or recommender algorithms.

In most of the book, we'll be talking about supervised algorithms, although we'll also talk about unsupervised tasks such as change detection and clustering.

The mapping function, f, predicts an outcome $f(X) \rightarrow \hat{Y}$. Each function is specified by a set of parameters, and the optimization results in a set of parameters that minimizes the mismatch between Y and $\hat{Y}$. Usually, this is done heuristically.

The match (mismatch) between Y and $\hat{Y}$ is measured by an error function, $g(\cdot)$, so the optimization consists of estimating parameters $\hat{\theta}$ for the function f that minimizes the error:

$$\hat{\theta} = \underset{\theta \in \Theta}{arg min} \; g(f_\theta(X), Y)$$

In this formula, since the error function is part of the optimization, $g(\cdot)$ is called the **objective function**.

In **reinforcement learning**, an agent is interacting with the environment through actions and gets feedback in the shape of rewards. You can find out more about reinforcement learning for time-series in *Chapter 11, Reinforcement Learning for Time-Series*.

Contrary to the situation in supervised learning, no labeled data is available, but rather the environment is explored and exploited based on the expectation of cumulative rewards.

Machine learning, the study of algorithms that improve with experience, can be traced back to the 1960s when statistical methods (discussed in *Chapter 1, Time-Series with Python*) were discovered. Let's start with a brief history of machine learning to give some context. This will provide some more terminology and a basic idea of the principal directions in machine learning. We'll give some more detailed context in the appropriate chapters.

History of machine learning

Biological neural networks were conceptualized as a mathematical model by Warren McCulloch and Walter Pitts in 1943 in what was the foundation of artificial neural networks.

Frank Rosenblatt developed the so-called **perceptron** in 1958, which is a **fully connected feed-forward neural network** in today's terms. This schematic shows a perceptron with two input neurons and a single output neuron (based on an image on Wikimedia Commons):

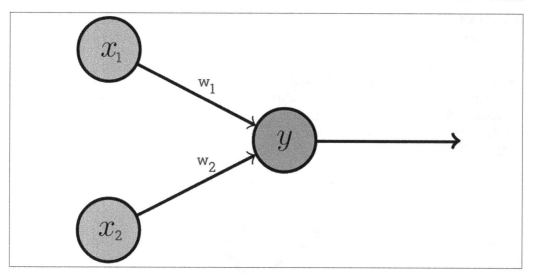

Figure 4.3: Perceptron

The connections to the output neuron y have weights w_1 and w_2. This is a simple linear model.

Another important step was how errors can be propagated backward through the network. The basics of **backpropagation** were published by Henry J. Kelley shortly afterward (1960) as a mechanism for training these networks.

This research was dealt a severe blow, however, when, in 1969, Marvin Minsky and Seymour Papert published the book "*Perceptrons*", which included a simple proof that linear functions (as in the 2-layer perceptron) could not model non-linear functions. According to the authors, this meant that perceptrons, wouldn't be useful or interesting in practice. The fact that perceptrons could have more than two layers, parameters of which could be learned via backpropagation, was glossed over in the book. Research in artificial neural networks only picked up again in the 1980s.

The **nearest neighbor algorithm** was described by Evelyn Fix and Joseph Hodges in 1951, and then expanded in 1967 by Thomas Cover and Peter E. Hart. The nearest neighbor algorithm can be applied to both classification and regression. It works by retrieving the k most similar instances between a new data point, and all known instances in the dataset (k is a parameter). In the case of classification, the algorithm votes for the most frequent label; in the case of regression, it averages the labels.

Another important milestone was the development of the decision tree algorithm. The ID3 decision tree algorithm (Iterative Dichotomiser 3) was published by Ross Quinlan in a 1979 paper and is the precursor to decision trees used today.

The **CART** algorithm (**Classification And Regression Tree**) was published by Leo Breiman in 1984. The **C4.5** algorithm, a descendant of ID3, came out in 1992 (Ross Quinlan) and is regarded today as a landmark in machine learning.

What was revolutionary about the decision tree is that it consists of step functions that partition the feature space of the data points into pockets that have a similar outcome. While many machine learning algorithms struggle when there are many interactions to consider, decision trees thrive in these situations. The following diagram illustrates what a decision tree looks like:

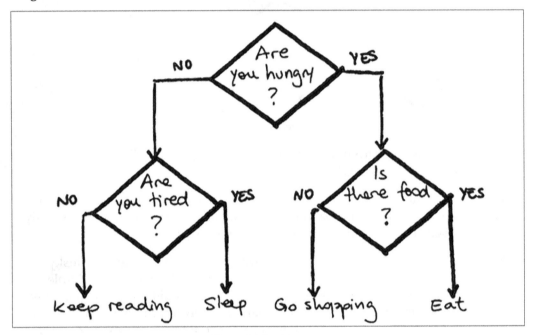

Figure 4.4: Decision tree

Each node or split in the tree is a single question based on the value of a feature. At each iteration during the tree construction, a statistical function called a split criterion is applied to decide on the best feature to query. Typical choices for a split criterion are the Gini impurity or information entropy, which both minimize the variability of the targets within branches.

Decision trees, in turn, form the basis of ensemble techniques such as the random forest or gradient boosted trees. There are two main ensemble techniques: boosting and bagging. **Boosting** was invented by Robert Schapire in 1990, and consists of incrementally adding base learners in a cascade. A **base learner** (also **weak learner**) is a very simple model that in itself is only weakly correlated to the targets. Each time when adding a new base learner to the existing ones, the importance (weights) of data points in the training set are rebalanced.

This means that in each iteration, the algorithm comes to grips with more and more samples that it struggles with, leading to higher precision with each new addition of a base learner.

This formed the basis for **AdaBoost**, an adaptive boosting algorithm, which won its inventors, Yoav Freund and Robert Schapire, the Gödel Prize, a prestigious recognition for outstanding papers in the area of theoretical computer science.

This illustration (from Wikipedia) shows how each base classifier is trained subsequently on different subsets of the dataset, where weights are changed for each new training:

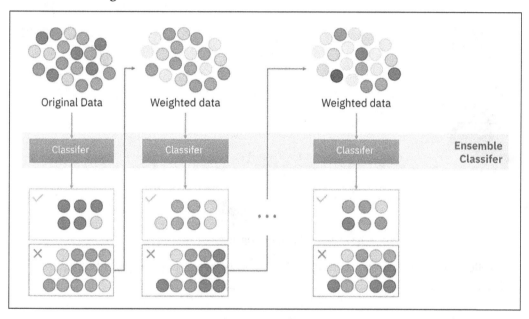

Figure 4.5: Boosting

Bagging is the basis for the random forest, and was invented in 1994 by Leo Breiman. Bagging consists of two parts, the bootstrap and aggregation. **Bootstrapping** is sampling with replacements from the training set. A separate model can be trained on each sample in isolation. These models together form an ensemble. The predictions from the individual models can then be aggregated into a combined decision, for example, by taking the mean.

The following diagram (source: Wikipedia) shows how a bagged ensemble is trained and used for prediction. This is how a **random forest** (Leo Breiman, 2001) learns with the decision tree as a base learner.

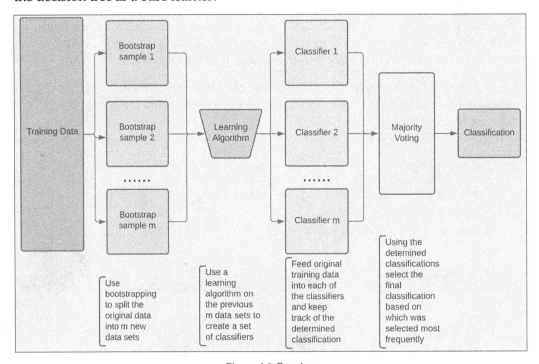

Figure 4.6: Bagging

The following table shows the main differences between bagging and boosting:

	Bagging	**Boosting**
Base learners are trained:	Independently (can be learned in parallel)	Sequentially
Weights are:	Left unchanged	Changed after every iteration
Base learners are weighted:	Equally	According to training performance

Figure 4.7: Differences between bagging and boosting

Gradient boosting (developed by Friedman and others) is a further extension of boosting with Unhyphenate. In gradient boosting, new weak learners are added in a fashion that they are maximally correlated with the negative gradient of the loss function. These are some popular implementations for gradient boosted trees:

- CatBoost (by Andrey Gulin and others at Yandex)
- Light Gradient Boosting Machine (LightGBM, at Microsoft)
- XGBoost

Backpropagation was rediscovered in 1986 by David Rumelhart, Geoffrey Hinton, and Ronald J. Williams. Shortly after, deeper networks were developed that could be applied to more interesting problems that attracted attention.

Between 1995 and 1997, Sepp Hochreiter and Jürgen Schmidhuber proposed a recurrent neural network architecture, the **long short-term memory** (**LSTM**). For many years, LSTMs constituted the state-of-the-art for many applications in voice recognition, translation, and more. Today, recurrent neural networks, have been largely replaced by transformers or ConvNets, even for sequence modeling tasks. With LSTM's high demands on computing resources, some people go as far as regarding the LSTM as obsolete given the alternatives.

Support Vector Machines (**SVMs**) were developed in the early 1990s at AT&T Bell Laboratories by Vladimir Vapnik and colleagues based on statistical learning frameworks described by Vapnik and Chervonenkis. In classification, SVMs maximize the distance between the two categories in a projected space. As part of the training, a hyperplane, called a support vector, is constructed that separates positive and negative examples.

In the next section, we'll go through the basics of machine learning modeling and scientific practices in model validation.

Machine learning workflow

In the next section, we'll go through the basics of time-series and machine learning.

Machine learning mostly deals with numerical data that is in tabular form as a matrix of size $N \times M$. The layout is generally in a way that each row $n \in \{1..N\}$ represents an observation, and each column $m \in \{1..M\}$ represents a feature.

In time-series problems, the column related to time doesn't necessarily serve as a feature, but rather as an index to slice and order the dataset. Time columns can, however, be transformed into features, as we'll see in *Chapter 3, Preprocessing time-series*.

Each observation is described by a vector of M features. Although a few machine learning algorithms can deal with non-numerical data internally, typically, each feature is either numerical or gets converted to numbers before feeding it into a machine learning algorithm. An example of a conversion is representing Male as 0 and Female as 1. Put simply, each feature can be defined as follows:

$$x \in X \subset \mathbb{R}^{\{N,M\}}$$

The machine learning workflow can be separated into three processes, as shown in the following diagram. I've added data loading and time-series analysis, which informs machine learning.

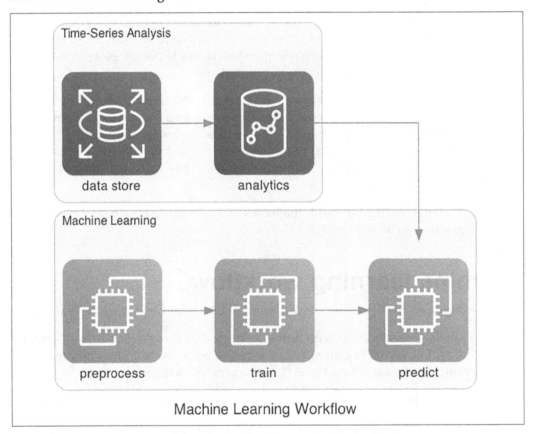

Figure 4.8: Machine learning workflow

We first must transform (or preprocess) our data, train or fit a model, and then we can apply the trained model to new data. This diagram, very simplistic perhaps, puts the focus on the three different stages of the machine learning process. Each stage comes with its own challenges and particularities for time-series data.

This can also help to think about the data flow from input to transform to training to prediction. We should keep in mind the available historical data and its limitations, as well as the future data points that are to be used for predictions.

In the next section, we'll discuss the general principles of cross-validation.

Cross-validation

Here's a well-known saying in machine learning attributed to George Box, whom we've encountered several times already in this book: "All models are wrong, but some are useful."

Machine learning algorithms make repeatable decisions and, given the correct controls, these decisions can be free from the cognitive biases that underlie much of human decision making. The point is to make sure that our model is useful by validating performance. In machine learning, the process of testing a model on data it hasn't seen in training is called cross-validation (sometimes, **out-of-sample testing**).

To ensure that parameters estimated on a dataset of limited size are still valid for more data, we must go through a validation that makes sure that the quality holds up. For validation, we usually split the dataset into at least two parts, the training set and the test set. We estimate parameters on a training set, and then run the model on the test set to get an idea of the quality of the model on unseen data points. This is illustrated in the following diagram:

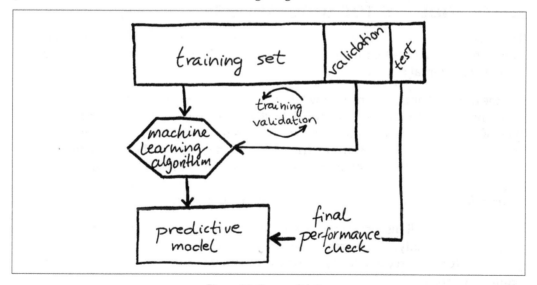

Figure 4.9: Cross-validation

Usually, in machine learning we would shuffle points randomly before splitting between training and test. However, in time-series, we would take older data points for training and newer points for testing. For instance, having 1 year of data available on the email opening propensities of customers, we would train a model on 9 months' worth of data, validate our model on 2 months' worth of data, and test the final performance on the dataset.

The use of validation and test can be seen as a nested process in the sense that the test set checks the main testing process that involves the validation dataset. Often, the separation of validation and test sets is omitted, so the dataset is split only into training and test sets.

A note about terminology: while a **loss function** is part of the optimization for training your model, a **metric** is used to evaluate your model. The evaluation can be post-hoc, after training, or during training as additional information. In this section, we'll discuss both metrics and loss functions.

It is good practice to start a project with an assessment of how to measure performance. We need to choose how to measure performance to translate the business problem into a metric or a loss. Some algorithms allow flexibility in the choice of objective functions, others don't, but we can measure performance with a different metric.

Next, we'll be discussing error and loss metrics for regression and classification.

Error metrics for time-series

Time-series data is defined as a set of data points containing details about different points in time. Generally, time-series data contains data points sampled or observed at an equal interval of time.

For the different applications that we discussed earlier, we need to be able to quantify the performance of the model, be it a regression, classification, or another type of model, and choose a metric that captures the performance we want to achieve. Once we have chosen a metric for our model, we can then build and train models to improve them. Often, we'd start with a simpler model and then try to improve on the performance of this simpler model as a baseline. In the end, we want to find the model that is best according to our metric.

In this section, we'll discuss commonly used performance measures and their properties. Generally, for an error measure, the smaller the values, the better the prediction (or the forecast). In changing the parameters of our model, we want to reduce the error.

There's not just a single metric that's apt for the purpose of any arbitrary application or dataset. Depending on the dataset, you might have to search and try different error metrics and see which one best captures your objective. In some circumstances, you might even want to define your own metric.

Regression

Time-Series regression is the task of identifying patterns and signals in the features in relation to the behavior of time-series, for example, how skill improves with the time invested in practice.

During training, when your regression model gives a result on the training set, we can utilize a metric that compares the model output to the training set values, and during validation, we can calculate the same measure to know how good our regression predictions line up to the validation set targets. The error metric summarizes the difference between the values predicted by your machine learning model and the actual values.

If $f(x_t)$ is a prediction of the model for time step t, and the actual target value is y_t, intuitively, for a particular point, t, of our dataset, the **forecast error** (also **prediction error** or **residual**) is the difference between the actual values of the target and the values our model predicts:

$$e_t = y_t - f(x_t)$$

This compares the actual target Y to the predicted targets $\bar{Y} = f(X)$. According to this formula, the error is negative if the prediction is higher than the actual target value. The **sum of squares of the residuals** (SS, also **residual sum of squares**) ignores the direction of the error:

$$SS = \sum_t (y_t - f(x_t))^2 = \sum_t e_t^2$$

While both the residual and the squared residual could already be used to measure the performance of predictions over a time-series, they are not commonly used as a regression metric or loss.

Let's start with the most commonly used metric for regression: the **coefficient of determination**. This is a relatively simple formula based on a ratio of the sum of the squares of the residuals, SS, and the total sum of squares, TSS, a measure of the variability:

$$R^2 = 1 - \frac{SS}{TSS}$$

In this fraction, the nominator is the sum of the squares of the residuals, SS, the unexplained variance.

The denominator is TSS, the total sum of squares. This is defined as $\sum_{t=1}^{N}(y_t - \bar{y})^2$, where $\bar{y}$ is the mean of the series, $\bar{y} = \frac{1}{N}\sum_{t}^{N} y$. The total sum of squares represents the explained variance of the time-series.

Basically, we are measuring the summed squares of residuals in relation to the total variance of the time-series. This fraction, between 0 and 1, where 0 is best – no error at all, is inverted by subtracting from 1, so that finally 0 is worst and 1 is best.

When expanded, this looks as follows:

$$R^2 = 1 - \frac{\sum_{t=1}^{N} e_t^2}{\sum_{t=1}^{N}(y_t - \bar{y})^2}$$

R^2 expresses the proportion of the variance in the dependent variable that is predictable from the independent variable. As mentioned, it is bounded between 0 and 1, where 1 means there's a perfect relationship, and 0 means there's no relationship at all.

The coefficient of determination, R^2, is not an error measure since an error measure expresses the distribution of residuals so that high is bad and low is good. We could, however, express an error measure, let's call it the **r-error (RE)**, very similar to the above, as:

$$RE = \frac{\sum_{t=1}^{N} e_t^2}{\sum_{t=1}^{N}(y_t - \bar{y})^2}$$

This is rarely used in practice. An error measure very similar to RE is the **mean relative absolute error (MRAE)**, which we'll discuss further ahead.

Naively, we could take the average error, where we just take the mean over the forecast error – the mean error:

$$ME = \frac{1}{N}\sum_{t=1}^{N} e_t,$$

Here, N is the number of points (or the number of discrete time steps). We calculate the error for each point and then take the mean over all these errors.

If the ME is positive, the model systematically underestimates the targets, if it's positive, it overestimates the targets on the whole. While this can be useful, it's a serious problem for an error metric, however, because the effects of positive and negative errors cancel each other out. Therefore, a low ME does not mean that predictions are good, rather that the average is close to zero.

Furthermore, most regression models include a constant term that is equal to the mean of the target, so this value would be exactly 0. In conclusion, our naïve measure is useless in practical settings.

I've included the ME for discussion of why most measures that are commonly used discard the direction of the error and for highlighting the importance of the main components of the basic error metrics:

- The residual operation
- The integration

In the case of the ME, the residual operation is the identify function, which means the residual doesn't change. More often, the square or absolute functions are used. The integration of the errors is often the (arithmetic) mean, but sometimes the median; however, it can be a more complex operation.

In practice, the most popular error metrics are the mean squared error (MSE), mean absolute error (MAE), and the root mean squared error (RMSE). These most important error metrics are defined in the following table:

Metric Name	Definition
Mean squared error	$$\text{MSE} = \frac{1}{N} \sum_{t=1}^{N} e_t^2$$
Mean absolute error	$$\text{MAE} = \frac{1}{N} \sum_{t=1}^{N} |e|_t$$
Root mean squared error	$$\text{RMSE} = \sqrt{\text{MSE}}$$

Figure 4.10: Popular regression metrics

With the **mean squared error (MSE)**, we calculate the residual for each point, then square them, so positive and negative errors don't cancel each other out. Then we take the mean over these squared errors. An MSE of 0 indicates perfect performance. This can happen with toy datasets that you can play around with for fun; however, in practice, this will only happen if you made a mistake in building your dataset or in validation, because real life is always more complex than you can capture with a model.

The **mean absolute error (MAE)** is very similar to the MSE, only instead of squaring the residuals, we take their absolute values. As opposed to the MSE, all errors contribute in linear proportion (rather than being squared).

A major difference between taking the absolute versus taking the square is in how outliers or extreme values are treated. The square function forces a higher weight on values that are very different. With the MSE, the error grows quadratically instead of linearly as is the case with the MAE. This means that the MSE punishes extreme values much more strongly and, as a result, it is less robust to outliers in the dataset than the MAE. The distribution of the errors is a major concern in choosing an error metric that's right for the job.

Another common metric is the **root mean squared error (RMSE)**, or **root mean square deviation (RMSD)**, which, as the name suggests, is the square root of the MSE. In that sense, RMSE is a scaled version of the MSE. Which one to take between the two is a presentation choice – both of them would lead to the same models.

What makes the RMSE interesting as a choice is that it comes in the same units and scale as the predicted variable, which makes it more intuitive. Finally, the RMSE is equivalent to the standard deviation or the error. This connection between standard deviation and the distribution of errors is quite meaningful, and you can summarize the error distribution with other measures such as the standard error or confidence interval (both of which we've discussed in *Chapter 2, Time-Series Analysis with Python*).

There are many more metrics and they all have their purpose. The following table sums up a few more popular error metrics in time-series modeling:

Metric Name	Definition
Median absolute error	$$\text{MdAE} = median(|e|_t)$$
Mean absolute percentage error	$$\text{MAPE} = \frac{1}{N}\sum_{t=1}^{N}\frac{|e_t|}{|y_t|}$$
Symmetric mean absolute percentage error	$$\text{SMAPE} = \frac{1}{N}\sum_{t=1}^{N}\frac{|e_t|}{(|y_t|+|f(x_t)|)/2}$$
Normalized mean squared error	$$\text{NMSE} = \frac{\text{MSE}}{\sigma^2}$$

Figure 4.11: More metrics for regression

The **median absolute error (MdAE)** is similar to the MAE. However, instead of the mean operation for integration, a different average, the median, is employed. Since the median is unaffected by values at the tails, this measure is even more robust than the MAE.

The **mean percentage error (MAPE)** is the mean average error normalized by the target. 0 represents a perfect model, and higher than 1 means the model's predictions are systematically higher than the targets. The MAPE doesn't have an upper bound. Additionally, since it deals with percentages in terms of the target (scaling or dividing by the targets), positive and negative residuals are treated differently. As a result, if the prediction is bigger than the target, the MAPE is higher than for the same error in the other direction. Therefore, depending on the sign of the residual, the MAPE is higher or lower!

The common choice for the denominator is the target; however, you can also scale by the mean of the prediction and target. This is called the **symmetric mean absolute percentage error (SMAPE)**. The SMAPE has not only a lower bound but also an upper bound, which makes the percentage much easier to interpret.

Scaling can have different benefits as well. If you want to compare models validated on different datasets, the measures presented before wouldn't be helpful. The split between training, validation, and test sets is randomized, so when you compare model performances, any of these measures would confound the effects of dataset variance in the validation set and the effect of the model performance itself.

Therefore, the **normalized mean squared error (NMSE)** is particularly intuitive as a presentation choice over the MSE because it scales the model's performance with the deviation. The NMSE normalizes the MSE obtained after dividing it by the target variance.

There are lots of other error measures. Some error measures compare predictions against predictions of a naïve model that returns the average target value.

The prediction performance of this naïve model is:

$$\hat{e} = \frac{1}{N} \sum_{t=1}^{N} (f(x_t) - \bar{y})^2$$

We can normalize prediction errors by dividing the prediction error by this naïve prediction error.

This way, we can define several other measures:

Metric Name	Definition
Mean relative absolute error	$$\text{MRAE} = \frac{\frac{1}{N} \sum_{t=1}^{N} |e|_t}{\frac{1}{N} \sum_{t=1}^{N} |\hat{e}|_t}$$
Median relative absolute error	$$\text{MdRAE} = \frac{median(|e|_t)}{median(|\hat{e}|_t)}$$
Relative root mean squared error	$$\text{RelRMSE} = \frac{\sqrt{\frac{1}{N} \sum_{t=1}^{N} e_t^{\,2}}}{\sqrt{\frac{1}{N} \sum_{t=1}^{N} \hat{e}_t^{\,2}}}$$

Figure 4.12: Normalized regression metrics

All these measures should be intuitive if you have the idea of the naïve model in mind and you want to compare the performance of the naïve model with the same error metric by dividing.

The **mean relative absolute error** (**MRAE**) is very similar to the coefficient of determination, with the only difference being that the MRAE takes the averages rather than the sums.

Another error is the **root mean squared logarithmic error** (**RMSLE**).

Metric Name	Definition
Root mean squared logarithmic error	$$\text{RMSLE} = \sqrt{\frac{1}{N}\sum_{t=1}^{N}(\log(f(x_t)+1)-\log(y_t+1))^2}$$

Figure 4.13: Root mean squared logarithmic error

In the case of the RMSLE, you take the log of the residuals as the basic operation. This is to avoid penalizing large differences in the error when both predicted and true values are very high values. Because of the inflection point of the logarithm at 1, the RMLSE has the unique property that it penalizes the underestimation of the actual value more severely than it does for overestimation. This can be useful when the distributions of errors don't follow a normal distribution similar to the scaling operations that we've discussed in *Chapter 3, Preprocessing Time-Series*.

We could extend the number of metrics if we take into account metrics based on entropy, such as Theil's Uncertainty. **Theil's U** is a normalized measure of the total prediction error. U is between 0 and 1, where 0 means a perfect fit. It is based on the concept of conditional entropy, and can also be used as a measure of uncertainty or even as a correlation measure in categorical-categorical cases.

As these headers say, the first two concentrate on quantifying the performance of models. The last section is useful for distance-based models, which are often used as a solid baseline for performance.

Let's switch over to error metrics for classification tasks.

Classification

Many metrics are specific to more binary classification (where there are exactly two classes), although some of them can be extended to the case of multi-class classification, where the number of classes is bigger than two.

In binary classification, we can contrast the prediction against the actual outcome in a **confusion matrix,** where predictions and actual outcomes are cross-tabulated like this:

		Actual outcome	
		false	true
Predicted outcome	false	true negative (TN)	false negative (FN)
	true	false positive (FP)	true positive (TP)

Figure 4.14: Confusion matrix

This is a crucial visualization for classification tasks, and many measures are based on summarizing this.

Two of the most important metrics for classification are precision and recall. **Recall** is the ratio of the number of correctly predicted positive instances across all positive instances. We can also state this in terms of the confusion matrix as follows:

$$Recall = \frac{TP}{TP + FN} = \frac{TP}{P}$$

Recall is also called the **true positive rate** or **sensitivity**. It focuses on the true predictions, ignoring the negative instances; however, we might also want to know how accurate the positive predictions are. This is **precision** defined as:

$$Precision = \frac{TP}{TP + FP}$$

We can visualize these two metrics as follows:

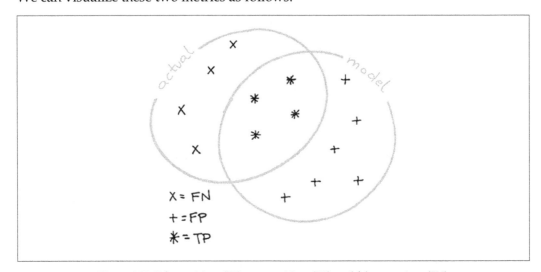

Figure 4.15: False positives (FP), true positives (TP), and false negatives (FN)

In this graph, **false positives (FP)**, **true positives (TP)**, and **false negatives (FN)** are shown. You can see instances that are actually true, instances that are classified as true by the model, and the intersection of the two – instances that are true and that the model classifies as true.

We can quickly count and calculate precision and recall. We have four true positives and six false positives. The precision for this example is therefore

$$\frac{TP}{TP + FP} = \frac{4}{4 + 6} = 0.4.$$

We have four false negatives. The recall is $\frac{TP}{FN + TP} = \frac{4}{4 + 4} = 0.5$.

Both recall and precision are obviously important, so why not integrate them? The F_1 score is the harmonic mean of precision and sensitivity:

$$F_1 = \frac{2\ recall \cdot precision}{recall + precision}$$

We can also parametrize the relative importance of recall and precision. This is a generalized version of the F_1 score, the F_β score:

$$F_\beta = \frac{(1 + \beta^2) \cdot recall \cdot precision}{\beta^2 \cdot recall + precision}$$

Another very useful metric comes from the **receiver operator curve (ROC)**, which plots the **true positive rate (TPR)** against the **false positive rate (FPR)** at various threshold settings. The **false positive rate**, also called the **false alarm ratio**, is defined analogously to the true positive rate (recall) as:

$$FPR = \frac{FP}{FP + TN} = \frac{FP}{F}$$

An ROC graph shows the relationship between sensitivity and specificity, with the general idea being that it is very hard to do both and find all positive instances (sensitivity) and to do them correctly. More often than not, you have to compromise between sensitivity and specificity. This plot illustrates how well your model maneuvers this issue. The **area under the curve** summarizes the plot and is a metric that's often used in practice.

Another less common metric is the **Correlation Ratio**, which was introduced by Karl Pearson as a measure of categorical-continuous association:

$$\eta = \sqrt{\frac{\sum_x n_x (\overline{y}x - \overline{y})^2}{\sum x, i (y_{xi} - \overline{y})^2}}$$

where n_x is the number of observations in category x, and we define:

$$\bar{y}x = \frac{\sum_i yxi}{n_x}, \bar{y} = \frac{\sum_i n_x \bar{y}_x}{\sum_x n_x}$$

The correlation ratio is based on the variance within individual categories and the variance across the whole population. η is in the range $[0,1]$ where 0 means a category is not associated, and 1 means a category is associated with absolute certainty.

In the next section, we'll examine similarity measures between time-series.

Comparing time-series

Similarity measures have applications in time-series indexing for retrieval in search, clustering, forecasting, regression, and classification, but if we want to decide whether two temporal sequences are similar, how do we measure the similarity?

The simplest would be to use the Pearson correlation coefficient; however, other measures can be more informative.

We'll go through a series of measures to compare a pair of time-series:

- Euclidean distance
- Dynamic time warping
- Granger causality

The **Euclidean distance**, a generic distance, is applicable to any pair of vectors, including time-series:

$$d(x,y) = \sqrt{\sum_t (x_t - y_t)^2}$$

The Euclidean distance can be useful; however, in practice, for time-series you can do better. You can take the Euclidean distance over the time-series that has been transformed by the fast Fourier transformed to a frequency domain.

Intuitively, the exact time position and its duration of events in time-series can vary. **Dynamic time warping** (DTW) is one of the algorithms for measuring similarity between two temporal sequences, which may vary in speed. Intuitively, the exact time position and its duration of events in time-series can vary. A similarity measure between time-series should be able to deal with these kinds of shifts and elongations.

In general, DTW is a method that calculates an optimal match between two given time sequences with certain restrictions and rules according to a heuristic. Basically, it attempts to match indexes from the first sequence to indexes from the other sequence. DTW is an edit distance – it expresses the cost of transforming a sequence t1 into t2.

DTW has been applied to automatic speech recognition because of its ability to cope with different speeds. DTW, however, fails at quantifying dissimilarity between non-matching sequences.

DTW is applied to each feature dimension independently and then the distances can be summed up. Alternatively, the warping can cover all features simultaneously by calculating the distance between two points as the Euclidean distance across all dimensions. Thus, this Dependent Warping (DTW_D) is a multivariate approach.

Granger causality determines if a time-series can help to forecast another time-series. Although the question of true causality in the measure is debatable, the measure considers values of one series prior in time to values of the other, and it can be argued that the measure shows a temporal relationship or a relationship in the predictive sense.

Granger causality is quite intuitive in both its idea and its formulation. Its two principles are (simplified):

1. The cause must precede the effect
2. The cause has a unique effect on the result

Therefore, if we can fit a model that shows that X and Y have a relationship in which Y systematically follows X, this is taken to mean that X Granger causes Y.

Machine learning algorithms for time-series

An important distinction in machine learning for time-series is the one between univariate and multivariate, in which algorithms are univariate, which means that they can only work with a single feature, or multi-variate, which means that they work with many features.

In univariate datasets, each case has a single series and a class label. Earlier models (classical modeling) focused on univariate datasets and applications. This is also reflected in the availability of datasets.

One of the most important repositories for time-series datasets, the **UCR** (**University of California, Riverside**) archive, which was released first in 2002, has provided a valuable resource for univariate time-series. It now contains about 120 datasets, but is lacking multivariate datasets. Furthermore, the M competitions (especially M3, 4, and 5) have a lot of available time-series datasets.

Multivariate time-series are datasets that have multiple feature dimensions. Many real-life datasets are inherently multivariate – multivariate cases are much more frequent in practice than univariate. Examples include human activity recognition, diagnoses based on an **electrocardiogram** (**ECG**), **electroencephalogram** (**EEG**), **magnetoencephalography** (**MEG**), and systems monitoring.

Only recently (Anthony Bagnall and others, 2018) created the **UAE** (**University of East Anglia**) archive with 30 multivariate datasets. Another archive for multivariate datasets is the MTS archive.

In the next section, we'll briefly discuss distance-based approaches.

Distance-based approaches

In the k-nearest-neighbor approaches (kNN for short), which we mentioned earlier, training examples are stored, and then, at inference time when a prediction for a new data point is required, the prediction is based on the closest k neighbors. This requires a distance measure between examples.

I've introduced two measures for time-series, **Dynamic Time Warping** (**DTW**) and Euclidean distances, earlier in this chapter. Many distance-based approaches take either of these as a distance measure.

Another approach that has been tried is extracting features from time-series, and then storing these extracted features for retrieval with kNN. These features include shapelets or **scale-invariant features** (**SIFT**). SIFT features are extracted from time-series as shapes surrounding the extremum (Adeline Bailly and others, 2015).

We've discussed shapelets and ROCKET in separate sections in *Chapter 3, Preprocessing Time-Series*, so we'll keep their descriptions brief, but focus on their applications in machine learning.

Shapelets

We've discussed shapelets in *Chapter 3, Preprocessing Time-Series*, so we'll keep it brief here. Shapelets for time-series were presented in the research paper "*Time-Series Shapelets: a novel technique that allows accurate, interpretable and fast classification*" (Lexiang Ye and Eamonn Keogh, 2011). The basic idea of shapelets is decomposing the time-series into discriminative subsections (called **Shapelets**). A few methods have been presented that are based on shapelet features.

The **Shapelet Transform Classifier** (**STC**; Hills and others, 2014) consists of taking the shapelets as feature transformation and then feeding the shapelets into a machine learning algorithm. They tested the C4.5 decision tree, Naïve Bayes, 1NN, SVM, and a rotation forest, but didn't find any significant differences between these methods in a classification setting.

The **Generalized random shapelet forest** (**gRFS**; Karlsson and others, 2016) follows the idea of the random forest. Each tree is built on a distinct set of shapelets of random length, which are extracted from one random dimension for each tree. A decision tree is trained on top of these shapelets. These random shapelet trees are then integrated as the ensemble model, which is the gRFS.

ROCKET

We've explained ROCKET in *Chapter 3, Preprocessing Time-Series*. Each input feature gets transformed separately by 10,000 random kernels (this number can be changed). In practice, this is a very fast process. These transformed features can be fed into a machine learning algorithm. Its inventors, Angus Dempster, François Petitjean, and Geoff Webb, recommended a linear model in the original publication (2019).

Recently, a new variant, MINIROCKET, was published that is about 75 times faster than ROCKET while maintaining roughly the same accuracy – *MINIROCKET: A Very Fast (Almost) Deterministic Transform for Time-Series Classification* (*Angus Dempster, Daniel F. Schmidt, and Geoff Webb, 2020*).

In machine learning research, **critical difference (CD) diagrams** are a powerful visualization tool for comparing outcomes of multiple algorithms. The average ranks indicate how algorithms stack up in relation to each other (a lower rank is better). Algorithmic results are compared statistically – a horizontal line links algorithms, where the differences between them can't be statistically separated.

Here's a critical difference diagram that illustrates the comparative performances of MiniRocket with other algorithms (from the MiniRocket repo by Dempster and others):

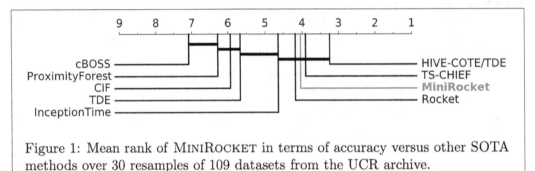

Figure 1: Mean rank of MINIROCKET in terms of accuracy versus other SOTA methods over 30 resamples of 109 datasets from the UCR archive.

Figure 4.16: Mean rank of MiniRocket in terms of accuracy versus other state-of-the-art approaches on 109 datasets of the UCR archive

The numbers show the rank of the algorithms across 109 datasets in the test. We can see that MiniRocket is better than Rocket, but worse than TS-CHIEF and HIVE-COTE, although the difference between them is not statistically significant.

We'll discuss InceptionTime in *Chapter 10, Deep Learning for Time-Series*. Some of the other methods mentioned will be introduced in the following sections.

Time-Series Forest and Canonical Interval Forest

The main innovation of the **Time-Series Forest** (**TSF**; by Houtao Deng and others, 2013) was the introduction of the entrance gain as a split criterion for the tree nodes. They showed that an ensemble classifier based on simple features such as mean, deviation, and slope outperforms 1NN classifiers with DTW while being computationally efficient (due to parallelism).

The **Proximity Forest (PF)**, introduced by a group of researchers lead by Geoff Webb, is a tree ensemble based on the similarity of each time-series to a set of reference time-series (distance-based features). They found that PF attains a classification performance comparable to BOSS and Shapelet Transforms.

TS-CHIEF, short for **Time-Series Combination of Heterogeneous and Integrated Embedding Forest**, comes from the same group (Ahmed Shifaz, Charlotte Pelletier, François Petitjean, and Geoff Webb, 2020), and it extends PF with dictionary-based (BOSS) and interval-based (RISE) splitters while keeping the original features introduced with PF. The authors claim that depending on the dataset size, it can run between 900 times and 46,000 times faster than HIVE-COTE.

The idea of the **Canonical Interval Forest** (CIF; by Matthew Middlehurst, James Large, and Anthony Bagnall, 2020) was to extend the TSF with the catch22 features. It is an ensemble of time-series trees based on the 22 Catch22 features and summary statistics extracted from phase-dependent intervals. They also used the entrance gain criterion for the trees.

In the next section, we describe the evolution of symbolic approaches, from BOSS to the **Temporal Dictionary Ensemble** (TDE).

Symbolic approaches

Symbolic approaches are methods that transform a numeric time-series to symbols from an alphabet.

Symbolic Aggregate ApproXimation (SAX) was first published by Eamonn Keogh and Jessica Lin in 2002. It extends **Piecewise Aggregate Approximation (PAA)**, which calculates averages within equal segments of the time-series. In SAX, these averages are then quantized (binned), so the alphabet corresponds to intervals of the original numerical values. The two important parameters are the number of segments in PAA and the number of bins.

The plot below (from Thach Le Nguyen's MrSEQL repository on GitHub) illustrates how SAX works:

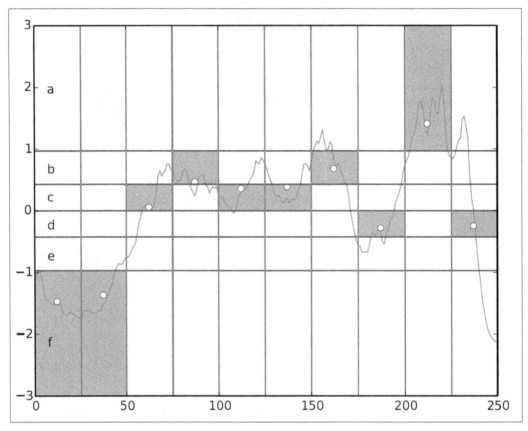

Figure 4.17: SAX

You can see the segments as a grid along the x axis and the bins as a grid along the y axis. Each segment is then replaced with its mean value. The time-series is discretized by replacing it in each segment with the bin ID (letter in the plot).

Symbolic Fourier Approximation (SFA; Patrick Schäfer and Mikael Högqvist, 2012) also transforms a time-series to a symbolic representation, but using the frequency domain. The dimensionality of the dataset is first reduced by performing Discrete Fourier Transformation, low-pass filtered, and then quantized.

The **Bag of SFA Symbols (BOSS;** Patrick Schäfer, 2015 and 2016) is based on histograms of n-grams to form a **bag-of-patterns (BoP)** from SFA representations. BOSS has been extended as **BOSS in Vector Space (BOSS VS)**. The BOSS VS classifier is one to four orders of magnitude faster than the state of the art and significantly more accurate than the 1-NN DTW.

Contract BOSS (cBOSS; Matthew Middlehurst, William Vickers, and Anthony Bagnall, 2019) speeds up BOSS by introducing a new parameter limiting the number of base models.

SEQL (Thach Le Nguyen, Severin Gsponer, and Georgiana Ifrim, 2017) is a symbolic sequence learning algorithm that selects the most discriminative subsequences for a linear model using greedy gradient descent. This is illustrated here (from the MrSEQL GitHub repo):

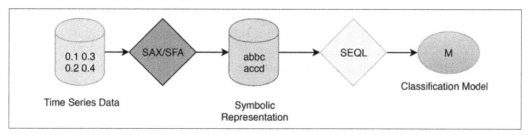

Figure 4.18: SEQL

The multiple representation sequence learner (**MrSEQL**; Thach Le Nguyen, Severin Gsponer, Iulia Ilie, and Georgiana Ifrim, 2019) is extending SEQL by selecting transformed features across multiple resolutions and multiple domains.

WEASEL+MUSE (Patrick Schäfer and Ulf Leser, 2017 and 2018) consists of two stages. WEASEL stands for **Word Extraction for Time-Series Classification**, while MUSE stands for **Multivariate Unsupervised Symbols and Derivatives**. This deserves emphasizing – while WEASEL is a univariate method, MUSE extends the method for multivariate problems.

In a first step, WEASEL derives features from windows at multiple lengths from the truncated Fourier transform and discretization. This acts in a way similar to a low-pass filter, keeping only the first l coefficients. These coefficients are then discretized into an alphabet of fixed size and counted as **Bag-of-Patterns** (**BOP**) in histograms. This is done in isolation for each feature.

In a second step (MUSE), the histogram features are concatenated across dimensions, and a statistical test, the $\chi 2$ test, is used for filter-based feature selection, resulting in a much smaller but more discriminative feature set.

These BOPs are then fed into a logistic regression algorithm for classification.

HIVE-COTE

The **Hierarchical Vote Collective of Transformation-Based Ensembles** (**HIVE-COTE**) is the current state of the art in terms of classification accuracy.

Proposed in 2016 and adapted in 2020 (Anthony Bagnall and others, 2020), it's an ensemble method that combines a heterogeneous collection of different methods:

- **Shapelet Transform Classifier (STC)**
- **Time-Series Forest (TSF)**
- **Contractable Bag of Symbolic-Fourier Approximation Symbols (CBOSS)**
- **Random Interval Spectral Ensemble (RISE)**

Random Interval Spectral Ensemble (RISE) is a tree-based time-series classification algorithm, originally introduced as **Random Interval Features (RIF)** at the same time as HIVE-COTE (Jason Lines, Sarah Taylor, and Anthony Bagnall, 2016). At each iteration of RISE, a set of Fourier, autocorrelation, and partial autocorrelation features are extracted, and a decision tree is trained. RISE's runtime complexity is quadratic to the series length, which can be a problem, and a new version has been released, **c-RISE** (*c* for *contract*), where the algorithm can be stopped earlier.

The runtime complexity of HIVE-COTE, the quadratic runtime to the length of the series, is one of the biggest obstacles to its adoption. STC and another model, the **elastic ensemble (EE)** were the two slowest base models in the original algorithm from 2016. One of the main differences of the new version (1.0) includes dropping EE. They re-implemented STC and BOSS to make them more efficient, and they replaced RISE with c-RISE.

Each of these base learners is trained separately. The base learners are weighted probabilistically based on a **Cross-Validation Accuracy Weighted Probabilistic Ensemble (CAWPE)** structure (James Large, Jason Lines, and Anthony Bagnall, 2019).

In publications postdating HIVE-COTE 1.0, the group showed that the ensemble is even stronger when replacing the CIF with the TSF (2020) and when replacing BOSS with the **Temporal Dictionary Ensemble (TDE, 2021)**.

In the next section, we will discuss the performance and trade-offs of different approaches.

Discussion

Generally, there's a trade-off between accuracy and prediction times, and in these methods, there's a huge difference in time complexity and model accuracy. This chart illustrates this compromise (from Patrick Schäfer's GitHub repository of SFA):

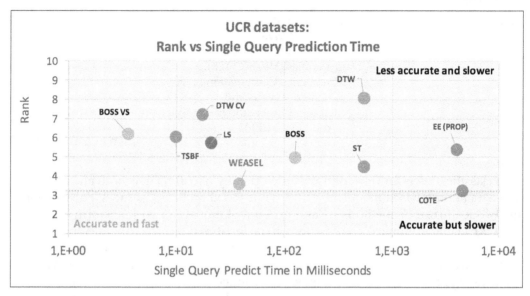

Figure 4.19: Machine learning algorithms: query time versus accuracy

Features could be the result of simple operations or themselves be the outcome of machine learning models. We could imagine second-order features as the combination of the original features, and third-order features as the combination of second-order features, and so on, a potentially large preprocessing pipeline, where features are combined and created.

We can sum up the different algorithms in this table:

Type	Univariate	Multivariate
Distance-based	DTW, Proximity Forest (PF)	DTW-D
Dictionary-based/Symbolic	BOSS, CBOSS, S-BOSS, WEASEL, Temporal Dictionary Ensemble (TDE), SAX-VSM, BOSS	WEASEL+MUSE
Shapelets	The Shapelet Transform Classifier (STC), MrSEQL	
Interval and Spectral-based	Time-Series Forest (TSF), Random Interval Spectral Ensemble (RISE)	
Deep learning	ResNet, FCN, InceptionTime	TapNet

Ensemble	The Hierarchical Vote Collective of Transformation-based Ensembles (HIVE-COTE), Time-Series Combination of Heterogeneous and Integrated Embeddings Forest (TS-CHIEF)	

Figure 4.20: Detailed taxonomy of time-series machine learning algorithms

This classification is far from perfect, but hopefully useful. TDE is both an ensemble and a dictionary-based model. HIVE-COTE is based on BOSS features. Furthermore, the two featurization methods – Random Convolutional Kernel Transform (ROCKET) and Canonical Time-Series Characteristics (Catch22), operate on features individually; however, machine learning algorithms that train on and predict based on these features as inputs can therefore work in a multivariate setting. The ROCKET features together with a linear classifier were indeed found to be highly competitive with multivariate approaches. Because of the high dimensionality, the machine learning model can potentially take interactions between the original features into account.

A review paper that I would highly recommend to readers is *"The great multivariate time-series classification bake-off"*, by Alejandro Pasos Ruiz, Michael Flynn, and Anthony Bagnall (2020). It compares state-of-the-art algorithms (16 of which were included in the analysis) on 26 multivariate datasets from the UAE archive. The approaches included the following:

- Dynamic time warping
- MUSE+WEASEL
- RISE
- CBOSS
- TSF
- gRSF
- ROCKET
- HIVE-COTE 1.0
- CIF
- ResNet
- STC

The critical difference diagram (as found on `timeseriesclassification.com`) shows the rank of the algorithms across 26 datasets in the test. Links between algorithms show that the differences between them can't be statistically separated (based on the Wilcoxon rank-sum test):

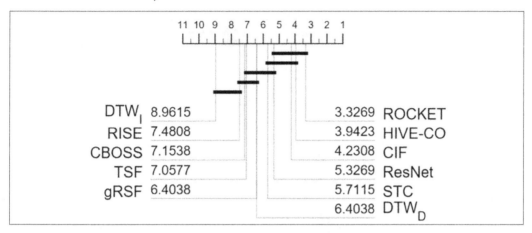

Figure 4.21 Critical difference diagram of time-series classification algorithms

They found a clique of top-performing classifiers, with ROCKET at the top achieving a considerable improvement in at least an order of magnitude less time. ROCKET was followed by HIVE-COTE and CIF.

In a study from 2019, Hassan Fawaz and others compared deep learning algorithms for time-series across 12 multivariate datasets from the MTSC archive. The fully connected convolutional network (FCN) was best, followed by ResNet – on 85 univariate datasets from the UCR repository, ResNet beat FCN to the top spot (winning on 50 out of 85 datasets). In a separate comparison involving just ResNet with some of the state-of-the-art non-deep learning methods on both univariate and multivariate datasets, they found that ResNet's performance was behind HIVE-COTE, although not significantly worse across datasets, while beating other approaches such as BOSS and 1NN with DTW (the latter to a statistically significant degree). We'll talk more about this paper in *Chapter 10, Deep Learning for Time-Series*.

In another comparison study on multivariate time-series classification on 20 datasets from the MTSC archive (Bhaskar Dhariyal, Thach Le Nguyen, Severin Gsponer, and Georgiana Ifrim, 2020), it was established that ROCKET won on 14 datasets and was much better than most deep learning algorithms, while at the same time being the fastest method – ROCKET's runtime on the 20 datasets was 34 minutes, while the DTW ran for days.

Here's the critical diagram created with Hassan Fawaz's Python script from their results:

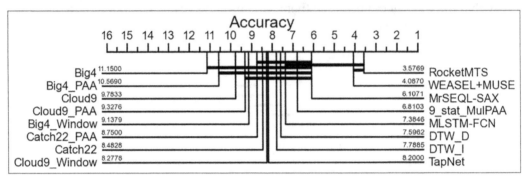

Figure 4.22: Critical difference diagram of multivariate time-series classification

Many different feature sets were tried, the best of which (9_stat_MulPAA) didn't end up far off.

Implementations

Great algorithms would be worth much less in practice without software that provides them in a way that makes them easy to use and reliable to use in a production setting of a company. Alternatively, implementing algorithms from scratch can take time, and is not without complications. Therefore, it's a boon that there are many reliable, available implementations in Python.

The following table summarizes implementations of supervised algorithms for regression and classification:

Algorithm	sktime	Pyts
Autoregressive Integrated Moving Average (ARIMA)	X	
DTW	X	X
BATS	X	
MUSE+WEASEL	X	X
MrSEQL	X	
ROCKET	X	X
BOSS	X	X
Bag-of-SFA Symbols in Vector Space (BOSSVS)		X

CBOSS	X	
SAX-VSM		X
RISE	X	
HIVE-COTE	X	
Time-Series Forest	X	

Figure 4.23: Pyts versus SkTime implementations of machine learning algorithms

It's not an accident that sktime has so many implementations. It is actively used in research activity by the group around Anthony Bagnall at the University of East Anglia. Pyts is being maintained by Johann Faouzi and Hicham Janati, postdoctoral fellows at the Paris Brain Institute and the **Centre de Mathématiques Appliquées (CMAP)** in Rémy. Johann Faouzi is also behind the tslearn library that implements time-series analysis and feature extraction algorithms.

I've omitted deep learning algorithms from the table, which are often implemented as part of different libraries. Please note that sktime allows use of the prophet forecaster through the same interface. For example, the sktime-DL library implements ResNet, InceptionTime, and TapNet algorithms, and dl-4-tsc implements more than a dozen deep learning models. We'll come to deep learning model implementations in *Chapter 10, Deep Learning for Time-Series*.

Facebook's Prophet contains a single model, a special case of the **Generalized Additive Model (GAM)**. The Statsmodels library contains a GAM as well as linear regression models and a **Generalized Linear Model (GLM)**, **moving average (MA)**, **Autoregressive Integrated Moving Average (ARIMA)**, and **Vector Autoregressions (VAR)**.

The Darts library provides a consistent interface to several models for time-series processing and forecasting. It includes both classical and deep learning algorithms:

- Exponential smoothing
- ARIMA
- Temporal convolutional network
- Transformer
- N-BEATS

This concludes our overview of time-series machine learning libraries in Python.

Summary

In this chapter, we've talked about both the context and the technical background of machine learning with time-series. Machine learning algorithms or models can make systematic, repeatable, validated decisions based on data. We explained the main machine learning problems with time-series such as forecasting, classification, regression, segmentation, and anomaly detection. We then reviewed the basics of machine learning as relevant to time-series, and we looked at the history and current uses of machine learning for time-series.

We discussed different types of methods based on the approach and features used. Furthermore, we discussed many algorithms, concentrating on state-of-the-art machine learning approaches.

I will discuss approaches including deep learning or classical models, such as autoregressive and moving averages, in chapters dedicated to them (for example, in *chapter 5, Time-Series Forecasting with Moving Averages and Autoregressive Models*, and *chapter 10, Deep Learning for Time-Series*).

5

Forecasting with Moving Averages and Autoregressive Models

This chapter is about time-series modeling based on moving averages and autoregression. This comprises a large set of models that are very popular in different disciplines, including econometrics and statistics. We'll discuss autoregression and moving averages models, along with others that combine these two, such as ARMA, ARIMA, VAR, GARCH, and others.

These models are still held in high esteem and find their applications. However, many new models have since sprung up that have been shown to be competitive or even outperform these simpler ones. Within their main application, however, in univariate forecasting, simple models often provide accurate or accurate enough predictions, so that these models constitute a mainstay in time-series modeling.

We're going to cover the following topics:

- What are classical models?
  - Moving average and autoregression
  - Model selection and order
  - Exponential smoothing
  - ARCH and GARCH
  - Vector autoregression

- Python libraries
  - statsmodels
- Python practice
  - Modeling in Python

We are going to start with an introduction to classical models.

What are classical models?

In this chapter, we'll deal with models that could be characterized as having a longer tradition, and are rooted in statistics and mathematics. They are used heavily in econometrics and statistics.

While there is considerable overlap between statistics and machine learning approaches, and each community has been absorbing the work of the other, there are still a few key differences. Whereas statistics papers are still overwhelmingly formal and deductive, machine learning researchers are more pragmatic, relying on the predictive accuracy of models.

We've talked about the very early history of time-series models in *Chapter 1, Introduction to Time-Series with Python*. In this chapter, we'll discuss moving averages and autoregressive approaches for forecasting. These were introduced in the early 20th century and popularized by George Box and Gwilym Jenkins in 1970 in their book *"Time-Series Analysis Forecasting and Control."* Crucially, in their book, Box and Jenkins formalized the ARIMA model and described how to apply it to time-series forecasting.

Many time-series exhibit trends and seasonality, while many of the models in this chapter assume stationarity. If a time-series is stationary, its mean and standard deviation stays constant over time. This implies that the time-series has no trend and no cyclic variability.

Therefore, the removal of irregular components, trends, and seasonal fluctuations is an intrinsic aspect of applying these models. The models then forecast what's left after removing seasonality and trend: business cycles.

Thus, to apply classical models, a time-series usually should be decomposed into different components. Thus, classical models are usually applied as follows:

1. Test for stationarity
2. Differencing [if stationarity detected]
3. Fit method and forecast
4. Add back the trend and seasonality

Most of the approaches in this chapter are relevant only to univariate time-series. Although extensions to multivariate time-series have been proposed, they are not as popular as the univariate versions. Univariate time-series consist of a single vector or, in other words, one value that changes over time. We'll see **Vector Autoregression (VAR)** at the end of the chapter though, which is an extension to multivariate time-series.

Another important consideration is that most classical models are linear, which means they assume linearity in the dependencies between values at the same time and between values at different time steps. In practice, the models in this chapter work well with a range of time-series that are stationary. This means that the distribution is the same over time. Examples of this are temperature changes over time. These models are especially valuable in a context where the amount of data available is small so that the extra estimation error in non-linear models dominates any potential gains in terms of accuracy.

However, the stationarity assumption implies that the application of the models in this chapter is restricted to time-series that have this property. Alternatively, we'd have to preprocess our time-series to enforce stationarity. In contrast, the development of statistical methods for nonlinear time-series analysis and forecasting has found much less prominence; however, there are some models such as the Threshold Autoregressive Model (which we won't cover here).

Finally, it is left to note that, while a reasonable first approach, many time-series such as temperatures can be predicted much more accurately by high-dimensional physics-based models of the atmosphere than by statistical models. This illustrates the point of complexity: essentially, modeling is condensing a set of hypotheses and formalizing them together with parameters.

Real-world time-series come from complicated processes that can be non-linear and non-stationary, and there's more than a single way of describing them, each of which has its up- and downsides. Thus, we can think of a modeling problem in terms of lots of parameters or just as a single or a couple of parameters. In a dedicated section below, we'll discuss the issue of selecting a model from a set of alternatives based on the number of parameters and the accuracy.

Nowadays, nonlinear models come from a different direction of research, either neural networks or the broader field of machine learning. We'll see neural networks in *Chapter 10, Deep Learning for Time-Series*, and we'll discuss and apply state-of-the-art machine learning in *Chapter 7, Machine Learning Models for Time-Series*.

Moving average and autoregression

Classical models can be grouped into families of models – **moving averages (MA)**, **autoregressive (AR)** models, ARMA, and ARIMA. These models were formalized and popularized over time in books and papers by many mathematicians and statisticians, including Peter Whittle (1951) and George Box and Gwilym Jenkins (1970). But let's start earlier.

The **moving average** marked the beginning of modern time-series predictions. In a moving average, the average (usually, the arithmetic mean) of values is taken over a specific number of time points (time frame) in the past.

More formally, the **simple moving average**, the unweighted mean over a period of k points, is formulated as:

$$\frac{x_1 + x_2 + \cdots + x_k}{k} = \frac{1}{k}\sum_{i=0}^{k} x_i,$$

where x_i represents the observed time-series.

The moving average can be used to smooth out a time-series, thereby removing noise and periodic fluctuations that occur in the short term, effectively working as a low-pass filter. Therefore, as mathematician Reginald Hooker pointed out in a publication in 1902, the moving average can serve to isolate trend and oscillatory components. He conceptualized trend as the direction in which a series is moving when oscillations are disregarded.

The moving average can smooth the trend and cycle over the history of a time-series; however, as a model, the moving average can be used to forecast into the future as well. The time-series is a linear regression of the current value of the series against observed values (error terms). The moving average model of order q, $MA(q)$, can be denoted as:

$$MA(q): x_t = \mu + \epsilon_t + \sum_{i=0}^{q}\varphi_i\epsilon_{t-i},$$

where μ is the average (expectation) of x_t (usually assumed to be 0), φ_i are parameters, and ϵ_i is random noise.

Educated in Cambridge, Hooker worked in the Statistical Branch of the Ministry of Agriculture, Fisheries, and Food of the United Kingdom. He was an out-of-hours statistician, writing about meteorology and socio-economic topics such as wages, marriage rates and trade, and crop forecasting.

The invention of **AR** techniques dates back to a paper by British statistician Udny Yule, a personal friend of Hooker's, in 1927 (*"On a Method of Investigating Periodicities in Disturbed Time-Series with special reference to Wolfer's Sunspot Numbers"*). An **autoregressive model** regresses the variable on its own lagged values. In other words, the current value of the value is driven by immediately preceding values using a linear combination.

Sunspot variations are highly cyclic as can be seen in this plot of sunspot observations over time (loaded through statsmodels data utilities):

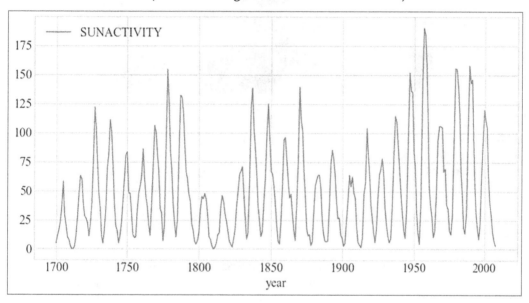

Figure 5.1: Sunspot observations by year

Yule formulized a linear model driven by noise as an application to sunspot numbers, the count of dark spots on the outer shell of the sun. The spots have their origin in giant explosions and indicate magnetic activity of the sun, and phenomena such as solar flares.

Here are two images of low and high solar activity according to sunspot numbers (from NASA):

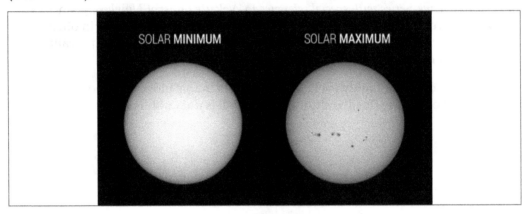

Figure 5.2: Solar activity

Today, we know that the solar cycle is a nearly periodic 11-year change in the solar magnetic activity between high magnetic activity (solar maximum) and low magnetic activity (solar minimum). At the high point, explosions (solar flares) can unleash charged particles into space, potentially endangering life on Earth.

Yule studied engineering at **University College London (UCL)**, went to work with Heinrich Hertz in Bonn, returned to UCL to work with Karl Pearson, and was later promoted to assistant professor at UCL. After first taking the post of statistics at UCL, he moved to Cambridge. He is remembered for his book on statistics "*An Introduction to the Theory of Statistics,*" which was first published in 1911 and went through many editions, as well as for his description of the process known today as preferential attachment, where the distribution of new quantities to nodes in a network is in accordance with how much the nodes already have; this is sometimes noted as "the richer get richer."

Andrey Kolmogorov defined the term **stationary process** in 1931, although Louis Bachelier had introduced a similar definition earlier (1900) using different terminology. **Stationarity** is defined by three characteristics:

1. Finite variation
2. Constant mean
3. Constant variation

Constant variation means that the variation of the time-series in a window between two points is constant over time: $\gamma_X(s, t) = \gamma_X(s + h, t + h)$, although it can change with the size of the window.

This is weak stationarity. In the literature, unless otherwise specified, usually stationarity means weak stationarity. Strict stationarity means that a time-series has a probability density function that is unchanged over time. In other words, under strict stationarity, the joint distribution over $(X_{t_1}, X_{t_2}, \dots, X_{t_k})$ is the same as over $(X_{t_1+h}, X_{t_2+h}, \dots, X_{t_k+h})$.

In 1938, Norwegian mathematician Herman Ole Andreas Wold described the decomposition of stationary time-series. He observed that stationary time-series can be expressed as the sum of a deterministic component (autoregressive) and a stochastic component (noise). This decomposition is termed after him today, as **Wold's decomposition**.

This leads to the formulation of the autoregressive model of order p, $AR(p)$, as:

$$AR(p): x_t = c + \sum_{i=1}^{p} \phi_i x_{t-i} + \epsilon_t,$$

where ϕ_i is a model parameter, c is a constant, and ϵ_t represents noise. In this equation, p is a measure of the autocorrelation between successive values of the time-series.

This work was later, in 1951, generalized to multivariate time-series in a Ph.D. thesis ("*Hypothesis Testing in Time-Series*") by New Zealander Peter Whittle, with Wold as his advisor. Peter Whittle is also credited with the integration of the AR and MA models into one, as the **autoregressive moving average** (**ARMA**). This was another milestone in the history of time-series modeling, bringing together the work of Yule and Hooker.

The ARMA model consists of two types of lagged values, one for the autoregressive component and the other for the moving average component. Therefore, we write $ARMA(p, q)$, with the first parameter p indicating the order of the autoregression, and the second, q, the order of the moving average, as:

$$ARMA(p, q): x_t = c + \epsilon_t + \sum_{i=1}^{p} \varphi_i x_{t-i} + \sum_{i=0}^{q} \phi_i \epsilon_{t-i}$$

ARMA assumes that the series is stationary. In practice, to ensure stationarity, preprocessing has to be applied.

The model parameters were estimated via the least-squares method until George Box and Gwilym Jenkins popularized their method of a maximum-likelihood estimation of the parameters.

George Box was one of the most influential figures of not only classical time-series prediction, but also the broader field of statistics. Drafted for World War II without having finished his studies in chemistry, he performed poison gas experiments for the army, teaching himself statistics for analysis in the process.

Once the war was over, he studied mathematics and statistics at the University College London, completing his Ph.D. with Egon Pearson, the son of Karl Pearson, as his advisor. He later headed a research group at Princeton, and then founded the statistics department at the University of Wisconsin–Madison.

Box and Jenkin's 1970 book "*Time-Series Analysis: Forecasting and Control*" outlined many applied examples for time-series forecasting and seasonal adjustment. The so-called Box-Jenkins method is one of the most popular forecasting methods. Their book also contained a description of the **autoregressive integrated moving average** model (**ARIMA**).

ARIMA(p, d, q) includes a data preprocessing step, called **integration**, to make the time-series stationary, which is by replacing values by subtracting the immediate past values, a transformation called **differencing**.

The model integration is parametrized by d, which is the number of times differences have been taken between current and previous values. As mentioned, the three parameters stand for the three parts of the model.

There are some special cases; ARIMA(p,0,0) stands for AR(p), ARIMA(0,d,0) for I(d), and ARIMA(0,0,q) is MA(q). I(0) is sometimes used as a convention to refer to stationary time-series, which don't require any differencing to be stationary.

While ARIMA type models effectively consider stationary processes, the **Seasonal Auto Regressive Integrative Moving Average** models (**SARIMA**), developed as an extension of the ARMA model, can describe processes that exhibit non-stationary behaviors both within and across seasons.

Seasonal ARIMA models are usually stated as ARIMA(p,d,q)(P,D,Q)m. The parameters deserve more explanation:

- m denotes the number of periods in a season
- P, D, Q parametrize the autoregressive, integration, and moving average components of the seasonal part
- p, d, q refer to the ARIMA terms, which we've discussed previously

P is a measure of autocorrelation between successive seasonal components of the time-series.

We can write out the seasonal parts to make this clearer. **Seasonal Autoregression, SAR**, can be stated as:

$$SAR(P): x_t = c + \sum_{i=1}^{P} \phi_i x_{t-s \cdot P} + \epsilon_t,$$

where s is the length of the seasonality.

Similarly, the **seasonal moving average, SMA**, can be written as follows:

$$\boldsymbol{SMA(Q)}: x_t = \mu + \epsilon_t + \sum_{i=0}^{Q} \varphi_i \epsilon_{t-s \cdot Q},$$

Please note that each of these components will use a distinct set of parameters.

For example, the model SARIMA(0,1,1)(0,1,1)12 process will contain a non-seasonal MA(1) term (with the corresponding parameter φ) and a seasonal MA(1) term (with the corresponding parameter ϕ).

Model selection and order

The parameter q in ARMA would typically be 3 or less, but this is more a reflection of computing resources rather than statistics. Today, to set the parameters p and q, we would typically look at the autocorrelation and partial autocorrelation plots, where we could see peaks in the correlation for each lag.

When we have different models, say models of different p and q, each trained on the same dataset, how do we know which one should we use? This is where model selection comes in.

Model selection is the methodology for deciding between competing models. One of the main ideas in model selection is Occam's razor, named after the English Franciscan friar and scholastic philosopher William of Ockham, who lived between circa 1287 and 1347.

According to Occam's razor, when choosing between competing solutions, one should prefer the explanation with the fewest assumptions. Ockham argued based on this idea that the principle of divine interventions is so simple that miracles are a parsimonious explanation. This rule, also called *"lex parsimoniae"* in Latin, expresses that a model should be parsimonious, which means that it should be simple yet have high explanatory power.

In science, simpler explanations are preferred out of the principle of falsifiability. The simpler a scientific explanation, the easier it can be tested, and possibly refuted – this lends the model scientific rigor.

ARMA and other models are usually estimated with the **maximum-likelihood estimation (MLE)**. In MLE, this means maximizing a likelihood function so that, given the parameters of the model, the observed data is most likely.

One of the most commonly used model selection criteria for the maximum-likelihood method is the **Akaike information criterion (AIC)**, after Hirotugu Akaike, who published it first in English in 1973.

AIC takes the log-likelihood l from the maximum-likelihood method and the number of parameters k in the model.

$$AIC = 2k - 2l$$

This is saying that the AIC equals two times the number of parameters minus two times the log-likelihood. In model selection, we would prefer the model with the lowest AIC, which means it has few parameters, but also high log-likelihood.

For an ARIMA model, we could write more specifically:

$$AIC = 2(p + q) - 2l$$

I've omitted the parameter d since it doesn't introduce additional estimations.

The **Bayesian Information Criterion (BIC)** was proposed for model selection a few years later (1978), by Gideon Schwarz, and looks very much like AIC. It additionally takes N, the number of samples in the dataset:

$$BIC = k \cdot \ln{(N)} - 2l,$$

According to BIC, we want a model with few parameters and high log-likelihood, but also a small number of training examples.

Exponential smoothing

Exponential smoothing, dating back to the work of Siméon Poisson, is a technique for smoothing time-series data using an exponential window function, which can be used to forecast time-series with seasonality and trend.

The simplest method, **simple exponential smoothing, SES,** s_t of a time-series x_t can be denoted as:

$$s_0 = x_0$$

$$s_t = \alpha x_t + (1 - \alpha)x_{t-1},$$

where α is the exponential smoothing factor (a value between 0 and 1).

Essentially, this is the weighted moving average with weights a and $1 - \alpha$. You can think of the second term, $(1 - \alpha)x_{t-1}$ as recursive, where, when expanding, $(1 - \alpha)$ gets multiplied by itself over and over – this is the exponential term.

Parameter α controls how much the smoothed value is determined by current versus previous values. The effect of this formula, as with moving averages, is that the result becomes smoother.

Interestingly, John Muth showed in 1960 that SES provided the optimal forecasts for a time-series where, at each time step, the values take a random step away from its previous value, and steps are independently and identically distributed in size plus noise. This kind of time-series is called a random walk, and sometimes, the price of a fluctuating stock is assumed to be following such a behavior.

The **Theta method**, another exponential smoothing method, is of particular interest to practitioners since it performed well in the M3 competition in 2000. The M3 competition, named after Spyros Makridakis, its organizer, who is a professor at the University of Nicosia and Director of the Institute for the Future, was a competition for forecasting across 3003 time-series from micro-economics, industry, finance, demographic, and other domains. One of its main conclusions was that very simple methods can perform well with univariate time-series forecasts. The M3 competition proved to be a watershed moment for forecasting, providing benchmarks and the **state of the art (SOTA)**, even though the SOTA has significantly changed since then, as we'll see in *Chapter 7, Machine Learning Models for Time-Series*.

The Theta method was proposed by Vassilis Assimakopoulos and Konstantinos Nikolopoulos in 2000 and re-stated in 2001 by Rob Hyndman and Baki Billah. The Theta model can be understood as **simple exponential smoothing (SES)** with drift.

This method is based on the decomposition of the de-seasonalized data into two lines. The first so-called "theta" line estimates the long-term component, the trend, and then takes the weighted average of this trend and the SES.

Let's state this more formally! The trend component is forecast like this:

$$T_t = c + b_0(t - 1) + \epsilon_t$$

In this equation, c is the intercept, b_0 is a coefficient multiplied by the time step, and ϵ_t is the residual. $c + b_0$ can be fit through ordinary least-squares.

The formula for Theta is then taking this trend and adding it to the SES as a weighted sum:

$$Theta: \hat{X}_t = (1 - \alpha)T_t + \alpha \hat{X}_{t-1}$$

Here, $\hat{x}_t$ is the forecast for X at time step t.

One of the most popular exponential smoothing methods is the **Holt-Winter's method**. Charles Holt, professor at the University of Chicago, first published a method for double exponential smoothing (1957) that allowed forecasting based on trend and level. His student Peter Winters extended the method to capture seasonality in 1960 (*"Forecasting sales by exponentially weighted moving averages"*). Even later, Holt-Winter's smoothing was extended to account for multiple seasonalities (n-order smoothing).

To apply the Holt-Winter's method, we first remove trend and seasonality. Then we forecast the time-series and add back the seasonality and trend.

We can distinguish additive and multiplicative variations of the method. Both trend and seasonality can be either additive or multiplicative.

An additive seasonality is seasonality added independently of the values of the series. A multiplicative seasonal component is added proportionally, when the seasonal effect decreases or increases with the values (or the trend) in the time-series. A visual inspection can help in deciding between the two variations.

The Holt-Winter's method is also called triple exponential smoothing because it applies exponential smoothing three times as we'll see. The Holt-Winter's method captures three components:

- An estimate of a level for each time point, L_t – this could be an average
- A trend component T
- Seasonality S_t with m seasons (the number of seasons in a year)

With additive trend and seasonality, in mathematical terms, the Holt-Winter's forecast for a value x_{t+k} is defined as:

$$HW: x_{t+k} = L_t + kT_t + S_{t+k-m}$$

With **multiplicative seasonality**, we multiply by the seasonality:

$$HW: x_{t+k} = (L_t + kT_t) \cdot S_{t+k-m}$$

The level is updated as follows:

$$L_t = \alpha \frac{x_t}{S_{t-m}} + (1 - \alpha)(L_{t-1} + T_{t-1})$$

We are updating the current level based on a weighted average of two terms, with α being the weight between them. These two terms are the previous level and the de-seasonalized value of the series.

In this equation, we are de-seasonalizing the series by dividing by the seasonality:

$$\frac{x_t}{S_{t-m}}$$

And the previous trend component gets added up to the previous level like this:

$$L_{t-1} + T_{t-1}$$

The trend update is as follows (for additive trend):

$$T_t = \beta(L_t - L_{t-1}) + (1 - \beta)T_{t-1}$$

Finally, the (multiplicative) seasonality update is as follows:

$$S_t = \gamma\frac{x_t}{L_t} + (1 - \gamma)S_{t-M}$$

We can switch these equations to additive variations as required. A more detailed treatment is beyond the scope of this book – we'll leave it here.

ARCH and GARCH

Robert F. Engle, professor of economics at MIT, proposed a model for time-series forecasting (1982) that he named **ARCH** (**Auto-Regressive Conditionally Heteroscedastic**).

For financial institutions, value at risk, the level of financial risk over a specific time period, is an important concept for risk management. Therefore, it is crucial to account for the covariance structure of asset returns. This is what ARCH does and explains its importance.

In fact, in recognition of his contributions to the field of time-series econometrics, Engle was awarded the 2003 Nobel Prize in Economics (Nobel Memorial Prize in Economic Sciences), together with Clive Granger, who we encountered earlier. The citation specifically mentioned his groundbreaking work on ARCH.

While, in ARMA-type models, returns are modeled as independent and identically distributed over time, ARCH allows for time-varying (heteroscedastic) error terms by parametrizing higher-order dependence between returns observed at varying frequencies.

In ARCH, the residuals are expressed as consisting of a stochastic, z_t, and a standard deviation, σ_t, both of which are time-dependent: $\epsilon_t = \sigma_t z_t$.

The standard deviation of the residual at time t is modeled depending on the residuals of the series at previous points:

$$\sigma_t^2 = \alpha_0 + \sum_{i=1}^{q} \alpha_i \epsilon_{t-i}^2,$$

where q is the number of preceding time points the variance depends on.

The model ARCH(q) can be determined using least-squares.

 The least-squares algorithm is to solve the linear equations $y = X \cdot \beta$ for β. It consists of finding the parameters that minimize the square of the error, $argmin_\beta \epsilon^2$.

GARCH (**generalized ARCH**) was born when Tim Bollerslev (1986) and Stephen Taylor (1986) both independently extended Engle's model to make it more general. The main difference between GARCH and ARCH is that the residuals come from an ARCH model rather than from an autoregressive model, AR.

Generally, before applying a GARCH or ARCH model, a statistical test for homoscedasticity is applied, in other words, whether the variance is constant over time. Commonly used is the ARCH-LM test, which works with the null hypothesis that the time-series has no ARCH effects.

Vector autoregression

All the previous forecasting methods presented in this chapter are for univariate time-series, that is, time-series that consist of a single time-dependent variable, a single vector. In practice, we usually know more than our single sequence of measurements.

For example, if our time-series is about the number of ice cream sales, we might know the temperatures or the sales of bathing suits. We could expect that the sales of ice cream are highly correlated to temperature, in fact, we might expect that ice cream is increasingly consumed when temperatures are high. Equally, we could speculate that the sales of bathing suits either coincide, pre-date, or post-date the sales of ice cream.

Vector autoregression models can track the relationships between several variables as they change over time. They can capture the linear dependence of a time-series on a vector of values that precedes the current timestamp, generalizing the AR model to multivariate time-series.

VAR models are characterized by their order, which refers to the number of preceding time points that go into the model. The simplest case, VAR(1), where the model takes just one lag of the series, can be formulated as follows:

$$VAR(1): x_t = c + \beta_1 x_{t-1} + \epsilon_t,$$

where c is a constant, the intercept of the line, β are the coefficients of the model, and ϵ_t is the error term at point t. x_t and c are vectors of length k, while β_1 is a $k \times k$ matrix.

A p-order model, VAR(p), is denoted as:

$$VAR(p): x_t = c + \sum_{i=0}^{p} \beta_i x_{t-i} + \epsilon_t,$$

VAR assumes that the error terms have a mean of 0 and that there is no serial correlation of error terms.

Just like vector autoregression is a multivariate generalization of autoregression, **vector ARIMA (VARIMA)** is an extension of the univariate ARIMA model to multivariate time-series. Although it was formalized already as early as 1957, available software implementations only appeared much later.

In the next section, we'll look at a few libraries in Python that we can use for forecasting with classical models.

Python libraries

There are a few popular libraries for classical time-series modeling in Python, but the most popular by far is statsmodels. The following chart compares the popularity of libraries in terms of the number of stars on GitHub:

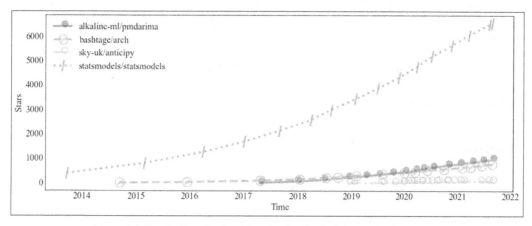

Figure 5.3: Popularity of Python libraries for classical time-series forecasting

Statsmodels is clearly the most popular among these libraries. I've only chosen to include libraries that are actively maintained and that implement the algorithms directly rather than importing them from other libraries. The SkTime or Darts libraries, for example, offer traditional forecasting models, but they are not implemented there, but in statsmodels.

pmdarima (originally pyramid-arima) contains a parameter search to help fit the best ARIMA model to univariate time-series. Anticipy contains a number of models, such as exponential decay and step models. Arch implements tools for financial econometrics and functionality for **Autoregressive Conditional Heteroscedasticity (ARCH)**.

While not as active as Scikit-Learn and only maintained by a couple of people, statsmodels is the go-to library for traditional statistics and econometrics approaches to time-series, with a much stronger emphasis on parameter estimation and statistical testing than machine learning.

Statsmodels

The statsmodels library can help to estimate statistical models and perform statistical tests. It's built on SciPy and NumPy and has lots of statistical functions and models.

The following table illustrates some of the modeling classes relevant to this chapter:

Class	Description
ar_model.AutoReg	Univariate Autoregression Model
arima.model.ARIMA	Autoregressive Integrated Moving Average (ARIMA) model
ExponentialSmoothing	Holt Winter's Exponential Smoothing
SimpleExpSmoothing	Simple Exponential Smoothing

Figure 5.4: A few models implemented in statsmodels

The ARIMA class also has functionality for SARIMA through a *seasonal_order* parameter, ARIMA, with seasonal components. By definition, ARIMA also supports MA, AR, and differencing (integration).

There are some other models, such as MarkovAutoregression, but we won't go through all of these – we will work through a selection.

Some other useful functions are listed here:

Function	Description
stattools.kpss	Kwiatkowski-Phillips-Schmidt-Shin test for stationarity
stattools.adfuller	Augmented Dickey-Fuller unit root test
stattools.ccf	The cross-correlation function
stattools.pacf	Partial autocorrelation estimate
stats.diagnostic.het_arch	Engle's Test for Autoregressive Conditional Heteroscedasticity (ARCH), also referred to as the ARCH-LM test
stattools.q_stat	Ljung-Box Q Statistic
tsa.seasonal.seasonal_ decompose	Seasonal decomposition using moving averages
tsa.tsatools.detrend	Detrend a vector

Figure 5.5: Useful functions in statsmodels

As a convention, we import statsmodels like this:

```
import statsmodels.api as sm
```

These statsmodels algorithms are also available through SkTime, which makes them available through an interface similar to the Sklearn interface.

This should be enough for a brief overview. Let's get into the modeling itself!

Python practice

As mentioned in the introduction to this chapter, we are going to be using the statsmodels library for modeling.

Requirements

In this chapter, we'll use several libraries, which we can quickly install from the terminal (or similarly from the anaconda navigator):

```
pip install statsmodels pandas_datareader
```

We'll execute the commands from the Python (or IPython) terminal, but equally, we could execute them from a Jupyter notebook (or a different environment).

Let's get down to modeling!

Modeling in Python

We'll work with a stock ticker dataset from Yahoo finance that we'll download through the yfinance library. We'll first load the dataset, do some quick exploration, and then we'll build several models mentioned in this chapter.

We'll load a series of Standard & Poor's depositary receipts (SPDR S&P 500 ETF Trust):

```
from datetime import datetime
import yfinance as yf

start_date = datetime(2005, 1, 1)
end_date = datetime(2021, 1, 1)

df = yf.download(
    'SPY',
    start=start_date,
    end = end_date
)
```

We have to indicate the date range and the ticker symbol. The daily prices come for Open, Close, and others. We'll work with the Open prices.

The index column is already a pandas DateTimeIndex so we don't have to convert it. Let's plot the series!

```
import matplotlib.pyplot as plt
plt.title('Opening Prices between {} and {}'.format(
    start_date.date().isoformat(),
    end_date.date().isoformat()
))
df['Open'].plot()
plt.ylabel('Price')
plt.xlabel('Date');
```

This gives us the following graph:

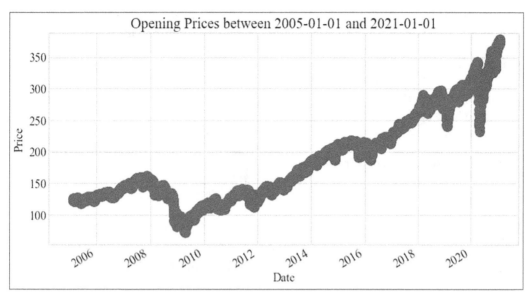

Figure 5.6: Standard & Poor's depositary receipt prices over time

Since this is daily data, and there are either 253 or 252 working days in the year, I've decided to resample the data to weekly data and make each year consistent.

```
df1 = df.reset_index().resample('W', on="Date")['Open'].mean()
df1 = df1[df1.index.week < 53]
```

Some years have 53 weeks. We can't handle that, so we'll get rid of the 53rd week. We now have weekly data over 52 weeks across 16 years.

One final fix: statsmodels can use the frequency information associated with the DateTimeIndex; however, this is often not set and df1.index.freq is None. So, we'll set it ourselves:

```
df1 = df1.asfreq('W').fillna(method='ffill')
```

If we check now, df1.index.freq is <Week: weekday=6>.

Setting the frequency can lead to missing values. Therefore, we are carrying over from the last valid value for missing values with the fillna() operation. If we don't do this, some of the models won't converge and give us NaN (not a number) values back instead of forecasts.

Now we need to get some idea of reasonable ranges for the order of the model. We'll look at the autocorrelation and partial autocorrelation functions for this:

```
import statsmodels.api as sm
fig, axs = plt.subplots(2)
fig.tight_layout()
sm.graphics.tsa.plot_pacf(df1, lags=20, ax=axs[0])
sm.graphics.tsa.plot_acf(df1, lags=20, ax=axs[1])
```

This gives us the following graph:

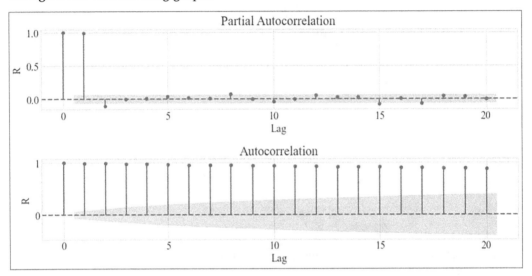

Figure 5.7: Partial autocorrelation and autocorrelation

These graphs show the correlation of the time-series with itself at lags of up to 20 time steps. R or ρ values close to 0 mean that consecutive observations at the lags are not correlated with one another. Inversely, correlations close to 1 or -1 indicate that there exists a strong positive or negative correlation between these observations at the lags.

Both the autocorrelation and the partial autocorrelation return confidence intervals. The correlation is significant if it goes beyond the confidence interval (represented as shaded regions).

We can see that the partial autocorrelation with lag 1 is very high and much lower for higher lags. The autocorrelation is significant and high for all lags, but the significance drops as the lag increases.

Let's move on to the autoregressive model. From here on, we'll use the statsmodels modeling functionality. The interface is very convenient, as you'll see.

We can't use an autoregressive model straight off because it needs the time-series to be stationary, which means the mean and variance is constant over time – no seasonality, no trend.

We can use statsmodels utilities to look at seasonality and trend from the time-series:

```
from statsmodels.tsa.seasonal import seasonal_decompose
result = seasonal_decompose(df, model='additive', period=52)
result.plot()
```

We set the period to 1 because each data point (row) corresponds to a year.

Let's see what the components look like:

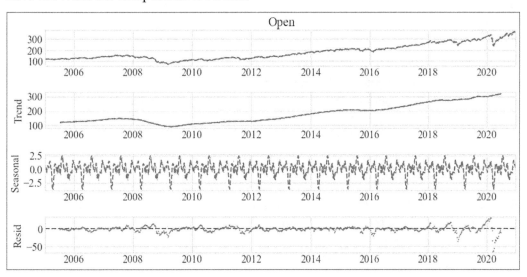

Figure 5.8: Seasonal decomposition of the time-series

The first subplot is the original time-series. There are both seasonality and trend in this dataset, which you can see separated out in the subplots.

As discussed, we need a stationary series for modeling. To establish stationarity, we need to remove the seasonal and trend components. We could also take off the seasonal or trend components that we estimated just before. Alternatively, we can use wrapper functionality in statsmodels or set the d parameter in ARIMA.

We can use the Augmented Dickey-Fuller and KPSS tests to check for stationarity:

```
from arch.unitroot import KPSS, ADF
ADF(df1)
```

We could have used `statsmodels.tsa.stattools.adfuller` or `statsmodels.tsa.stattools.kpss`, but we prefer the convenience of the ARCH library versions. We are leaving it to the user to check the output of the KPSS test. We get the following output:

```
            Augmented Dickey-
              Fuller Results
Test Statistic   1.344

      P-value   0.997

         Lags     9




Trend: Constant
Critical Values: -3.44 (1%), -2.87 (5%), -2.57 (10%)
Null Hypothesis: The process contains a unit root.
Alternative Hypothesis: The process is weakly stationary.
```

Figure 5.9: Output from the KPSS test for stationarity

Given the p-value of 0.997, we can reject our null hypothesis of the unit root, and we conclude that our process is weakly stationary.

So how do we find good values for the differencing? We can use the pmdarima library for this, where there is a function for precisely this purpose:

```
from pmdarima.arima.utils import ndiffs
# ADF Test:
ndiffs(df1, test='adf')
```

We get a value of 1. We would get the same values for the KPSS and the PP tests. This means that we can work off the first difference.

Let's start with an autoregressive model.

As a reminder, ARIMA is parametrized with parameters p, d, q, where:

- p is for the autoregressive model: AR(p)
- d is for the integration
- q is for the moving average: MA(q)

Therefore, ARIMA(p, d, 0) is AR(p) with a differencing of order d.

It is reassuring to know that statsmodels checks and warns if the stationarity assumption is not warranted. Let's try to run to fit the following AR model:

```
mod = sm.tsa.arima.ARIMA(endog=df, order=(1, 0, 0))
res = mod.fit()
print(res.summary())
```

```
UserWarning: Non-stationary starting autoregressive parameters found.
Using zeros as starting parameters.
  warn('Non-stationary starting autoregressive parameters'
```

Since we already know we need a differencing of one degree, we can set d to 1. Let's try again. This time, we'll use the STLForecast wrapper that removes seasonality and adds it back in. This is necessary since ARIMA can't handle seasonality out of the box:

```
from statsmodels.tsa.forecasting.stl import STLForecast
mod = STLForecast(
  df1, sm.tsa.arima.ARIMA,
  model_kwargs=dict(order=(1, 1, 0), trend="t")
)
res = mod.fit().model_result
print(res.summary())
```

We get this summary:

```
                               SARIMAX Results
==============================================================================
Dep. Variable:                     y   No. Observations:                  834
Model:                ARIMA(1, 1, 0)   Log Likelihood               -1965.555
Date:               Sun, 22 Aug 2021   AIC                           3937.110
Time:                       16:58:58   BIC                           3951.285
Sample:                   01-09-2005   HQIC                          3942.545
                        - 12-27-2020
Covariance Type:                 opg
==============================================================================
                 coef    std err          z      P>|z|      [0.025      0.975]
------------------------------------------------------------------------------
x1             0.2771      0.119      2.327      0.020       0.044       0.511
ar.L1          0.2502      0.022     11.402      0.000       0.207       0.293
sigma2         6.5618      0.184     35.641      0.000       6.201       6.923
===================================================================================
Ljung-Box (L1) (Q):                   0.24   Jarque-Bera (JB):               764.77
Prob(Q):                              0.62   Prob(JB):                         0.00
Heteroskedasticity (H):               3.21   Skew:                            -0.28
Prob(H) (two-sided):                  0.00   Kurtosis:                         7.66
===================================================================================

Warnings:
[1] Covariance matrix calculated using the outer product of gradients (complex-step).
```

Figure 5.10: Summary of our ARIMA model

This result summary gives all the key statistics. We see that the model was ARIMA(1, 1, 0). The log-likelihood was -1965. We also see the BIC and AIC values that we can use for model selection if we want.

Please note that we need to set trend="t" here so that the model includes a constant. If not, we would get a spurious regression.

How can we use this model? Let's do some forecasting!

```
STEPS = 20
forecasts_df = res.get_forecast(steps=STEPS).summary_frame()
```

This gives us a forecast 20 steps into the future.

Let's visualize this!

```
ax = df1.plot(figsize=(12, 6))
plt.ylabel('SPY')
forecasts_df['mean'].plot(style='k--')
ax.fill_between(
    forecasts_df.index,
    forecasts_df['mean_ci_lower'],
    forecasts_df['mean_ci_upper'],
    color='k',
    alpha=0.1
)
```

Here's what we get:

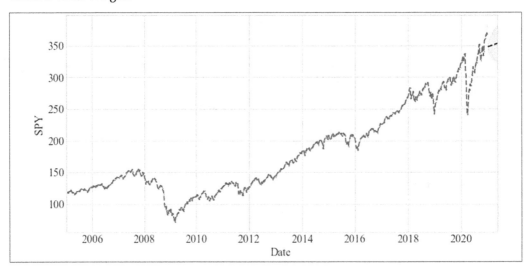

Figure 5.11: Forecast of prices for the SPY ticker symbol

The solid line is the data that we know. The dotted line represents our forecast 20 years into the future. The gray area around our forecast is the 95% confidence interval.

This doesn't look too bad. It's left as an exercise to the reader to try with different parameters. Interesting ones to change are the trend parameter and the order of the model.

For the moving average, let's create different models to see the difference in their forecasts!

First, we'll produce the forecasts:

```
forecasts = []
qs = []
for q in range(0, 30, 10):
    mod = STLForecast(
            df1, sm.tsa.arima.ARIMA,
            model_kwargs=dict(order=(0, 1, q), trend="t")
        )
    res = mod.fit()
    print(f"aic ({q}): {res.aic}")
    forecasts.append(
            res.get_forecast(steps=STEPS).summary_frame()['mean']
        )
    qs.append(q)

forecasts_df = pd.concat(forecasts, axis=1)
forecasts_df.columns = qs
```

In the loop, we are iterating over different q parameters, choosing 0, 10, and 20. We estimate moving average models with these values of q and forecast 20 years ahead. We also print the AIC values corresponding to each q. This is the output we get:

```
aic (0): 3989.0104184919096
aic (10): 3934.375909262983
aic (20): 3935.3355340835
```

Now, let's plot the three forecasts similar to how we did before:

```
ax = df1.plot()
plt.ylabel('SPY')
forecasts_df.plot(ax=ax)
```

Here's the new plot:

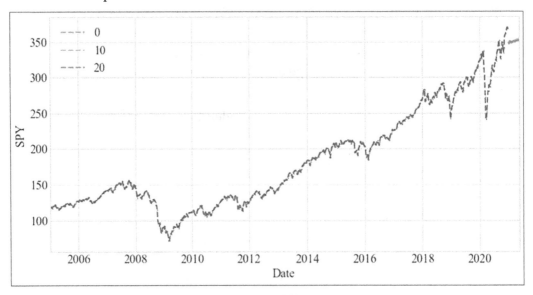

Figure 5.12: Forecasts using different q parameters

So, which one of these models is statistically better?

Let's get back to the AIC. The lower the AIC value, the better the model given its log-likelihood and the number of parameters.

In this case, the order of 10 gives us the lowest AIC, and according to this criterion, we should therefore choose q=10. Of course, we only tried three different values. I'll leave it as an exercise to come up with a more reasonable parameter value for q.

Please note that the pmdarima library has functionality for finding the optimal parameter values, and the SkTime library provides an implementation for automatic discovery of the optimal order of an ARIMA model: AutoARIMA.

Let's move on and make a forecast using an exponential smoothing model.

In the loop, we are iterating over different q parameters, choosing 0, 10, and 20. We estimate moving average models with these values of q and forecast 20 years ahead. We also print the AIC values corresponding to each q. This is what we get:

```
mod = sm.tsa.ExponentialSmoothing(
        endog=df1, trend='add'
    )
res = mod.fit()
```

This fits the model to our data.

Let's get the forecasts for the next 20 years:

```
forecasts = pd.Series(res.forecast(steps=STEPS))
```

Now, let's plot the forecasts:

```
ax = df.plot(figsize=(12, 6))
plt.ylabel('SPY')
forecasts.plot(style='k--')
```

Here's the plot:

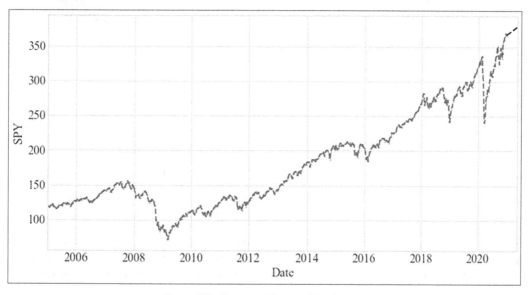

Figure 5.13: Exponential smoothing forecast

Until now, we've just looked at the graphs of 20 step-ahead forecasts. We still haven't been very sophisticated with an analysis of our model performance. Let's have a look at the errors!

For this, we first have to split our dataset into training and testing. We can do an n-step ahead forecast and check the error. We'll just take the time-series running up to a certain time point as the training data, and the time points after that as test data, where we can compare the predictions to actual data points:

```
from statsmodels.tsa.forecasting.theta import ThetaModel
train_length = int(len(df1) * 0.8)
tm = ThetaModel(df1[:train_length], method="auto",deseasonalize=True)
```

```
res = tm.fit()
forecasts = res.forecast(steps=len(df1)-train_length)
ax = df1.plot(figsize=(12, 6))
plt.ylabel('SPY')
forecasts.plot(style='k--')
```

Here's the plot:

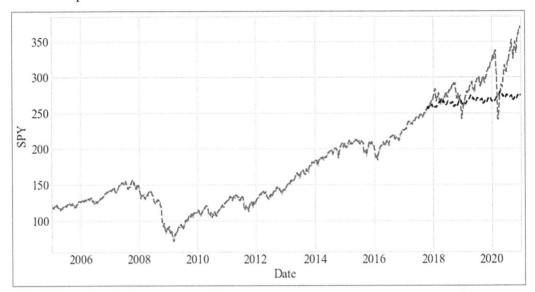

Figure 5.14: ThetaModel forecast

The dotted line is the prediction. It doesn't seem to line up well with the actual behavior of the time-series. Let's quantify this using one of the error metrics we've discussed in the previous chapter, *Introduction to Machine Learning for Time-Series*:

```
from sklearn import metrics
metrics.mean_squared_error(forecasts, df1[train_length:], squared=False)
```

We get a value of 37.06611385754943. This is the root mean squared error (we are setting the squared parameter to False).

In forecasting competitions, such as on the Kaggle website, the lowest error wins. In real life, parsimony (simplicity) is important as well; however, we usually still aim for the lowest error (by whichever chosen metric) that we can get.

There are a lot more models to explore and to play around with, but it's time to conclude this chapter.

Summary

In this chapter, we've talked about time-series forecasting based on moving averages and autoregression. This topic comprises a large set of models that are very popular in different disciplines, such as econometrics and statistics. These models constitute a mainstay in time-series modeling and provide state-of-the-art forecasts.

We've discussed autoregression and moving averages models, and others that combine these two, including ARMA, ARIMA, VAR, GARCH, and others. In the practice session, we've applied a few models to a dataset of stock ticker prices.

6

Unsupervised Methods for Time-Series

We've discussed forecasting in the previous chapter, and we'll talk about predictions from time-series in the next chapter. The performance of these predictive models is easily undermined by major changes in the data. Recognizing these changes is the domain of unsupervised learning.

In this chapter, we'll describe the specific challenges of unsupervised learning with time-series data. At the core of unsupervised learning is the extraction of structure from time-series, most importantly recognizing similarities between subsequences. This is the essence of anomaly detection (also: outlier detection), where we want to identify sequences that are notably different from the rest of the series.

Time-series data is usually non-stationary, non-linear, and dynamically evolving. An important challenge of working with time-series is recognizing the changes in the underlying processes. This is called change point detection (CPD) or drift detection. Data changes over time, and it is crucial to recognize how much it changes. This is worth diving into, because the existence of change points and anomalous points are common problems with real-world applications.

In this chapter, we'll concentrate on anomaly detection and CPD, while in *Chapter 8, Online Learning for Time-Series*, we'll look at drift detection in more detail. We'll start with an overview and definitions, before looking at industry practices at big tech companies.

We're going to cover the following topics:

- Unsupervised methods for time-series
- Anomaly detection
- Change detection
- Clustering
- Python practice

We'll start with a general introduction to unsupervised learning with time-series.

Unsupervised methods for time-series

The main difference between time-series and other types of data is the dependence on the temporal axis; a correlation structure at one point t_1 could have very different information to the same structure at point t_2. Time-series often contain lots of noise and have high dimensionality.

To reduce the noise and decrease the dimensionality, dimensionality reduction, wavelet analysis, or signal processing techniques such as filtering, for example, Fourier decomposition, can be applied. This is often at the basis of anomaly detection or CPD, the techniques we are discussing in this chapter. We'll discuss drift detection in *Chapter 8, Online Methods for Time-Series*.

We'll be talking in detail about anomalies and change points, and it might be helpful to see an illustration of how anomalies and change points can look like. In the article *"Social tipping dynamics for stabilizing Earth's climate by 2050"* by Ilona Otto and others, they analyzed whether and how a change in greenhouse gas emissions based on social dynamics could transform countries into carbon-neutral societies. They project global warming according to different scenarios in the following plot with a tipping point (another word for change point) around 2010 and the early 2020s (chart adapted from their article):

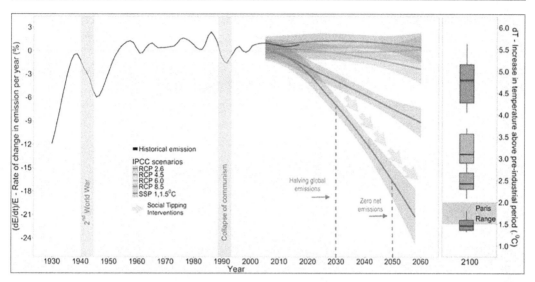

Figure 6.1: Possible change points based on greenhouse gas emissions

Global temperatures have cycled between periods of glaciations and warm periods, each cycle taking somewhere in the order of tens of thousands of years. For the last few thousand years, the climate was cooling leading to widespread speculations as late as the 1970s around a cooling trend that could lead to another ice age. However, data indicates that since the beginning of industrialization, largely driven by the burning of fossil fuels, global temperature has increased about a full 1°C.

Therefore, the beginning of the industrial period could be considered a change point for global temperatures as the following graph illustrates (source: Wikimedia Commons):

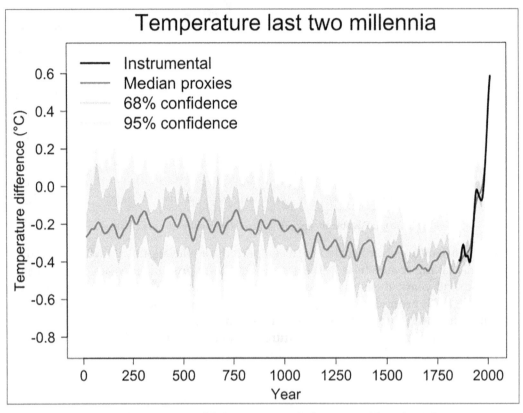

Figure 6.2: Change point in global temperatures: the beginning of the industrial age

In the graph above, the change point at the onset of the industrial revolution precedes the anomaly of modern temperature rises.

For a human, it's relatively easy to point out change points or anomalies, especially in hindsight when the full historical data is available. For automatic detection, there are lots of different ways to find salient points. In a practical context, it's important to carefully balance detection rates and false positives.

Anomaly detection

In anomaly detection, we want to identify sequences that are notably different from the rest of the series. Anomalies or outliers can sometimes be the result of measurement error or noise, but they could indicate changes to behavior or aberrant behavior in the system under observation, which could require urgent action.

An important application of anomaly detection is automatic real-time monitoring of potentially complex, high-dimensional datasets.

It's time for an attempt at a definition (after D.M. Hawkins, 1980, "*Identification of Outliers*"):

Definition: An outlier is a data point that deviates so significantly from other observations that it could have been generated by a different mechanism.

Let's start with a plot, so we can see how an anomaly might look graphically. This will also provide us context for our discussion.

Anomaly detection methods can be distinguished between univariate and multivariate methods. Parametric anomaly detection methods, by the choice of their distribution parameters (for example, the arithmetic mean), place an assumption on the underlying distribution – often the Gaussian distribution. These methods flag outliers, points that deviate from the model's assumptions.

In the simplest case, we can define an outlier as follows as the z-score of the observation x_i with respect to the distribution parameters:

$$ outlier(\hat{\mu}, \hat{\sigma}) = \left\{ x \colon \frac{|x_i - \hat{x}|}{\hat{\sigma}} > \epsilon \right\} $$

The z-score measures the distance of each point from the moving average or sample mean, $\hat{x}$, in units of the moving or sample standard deviation $\hat{\sigma}$. It is positive for values that lie above the mean, and negative for those that lie below the mean.

In this formula, $\hat{\mu}$ and $\hat{\sigma}$ are the estimated mean and standard deviation of the time-series and x is the point that we want to test. Finally, ϵ is a threshold dependent on the confidence interval that we are interested in – often, 2 or 1.96 are chosen for this, corresponding to a confidence interval of 95%. In this way, outliers are points that occur 5% or less of the time.

The z-score makes an assumption of normal-distributed data; however, the mean and standard deviation used in the outlier formula above can be replaced by other measures that do away with this assumption. Measures such as the median or the interquartile range (as discussed in *Chapter 2*, *Time-Series Analysis with Python*) are more robust to the distribution.

The Hampel filter (also: Hampel identifier) is a special case for this, where the median and the **median absolute deviation** (**MAD**) are employed:

$$ HF(x) = \left\{ x \colon \frac{x - median(x)}{1.4826\ MAD(x)} > \epsilon \right\} $$

In this equation, the sample mean is replaced by the (sample) median and the standard deviation by the MAD, which is defined as:

$$Median(|x_i - median(x)|)$$

The median, in turn, is the middle number in a sorted list of numbers.

In the Hampel filter, each observation, x, will be compared to the median. In the case of the normal distribution, the Hampel filter is equivalent to the z-score, and epsilon can be chosen the same way as for the z-score.

In the multivariate case, the outlier function can be expressed as the distance (or, inversely: similarity) to a point in the model distribution such as the center of gravity, the mean. For example, we could take the covariance of the new observation to the mean.

While these former methods are restricted to low-dimensional or univariate time-series, distance-based methods can handle much larger spaces. Distance-based anomaly detection methods effectively cluster points into different groups, where small groups will be labeled outliers. In these methods, the choice of distance measure is crucial.

Some of the challenges to detect anomalies in time-series are:

- Lack of a definition of outliers
- Noise within the input data
- Complexity of time-series
- High imbalance

We often don't really know what outliers look like. In practical settings, we often don't have labels for the outliers – rendering benchmarking against real cases impossible. As for the complexity, time-series change over time, they are often non-stationary and the dependence between variables can be non-linear. Finally, we typically have a lot more normal observations than outliers.

A requirement for deploying anomaly detection models as services at scale is that they should be able to detect anomalies in real time.

Applications for anomaly detection encompass the ones in this diagram:

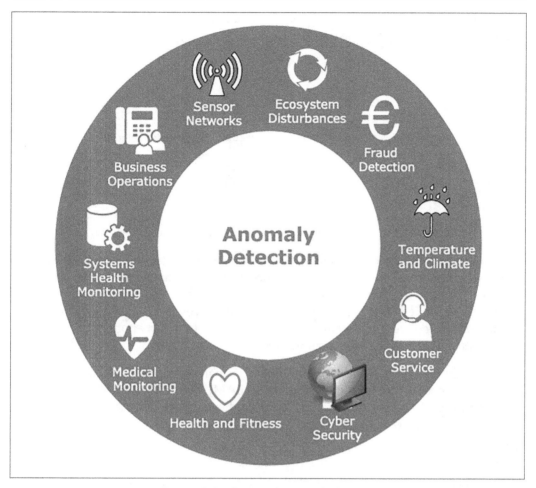

Figure 6.3: Applications for anomaly detection

Some examples could be fraud detection with payments, network security (cyber intrusion), medical monitoring, or sensor networks. In medical monitoring, we want to work with real-time monitoring of physiological variables including heart rate, electroencephalogram, and electrocardiogram for alerting in case of acute emergencies. Anomaly alerts in sensor networks can help prevent cases of industrial damage, for example.

This diagram illustrates the main types of anomaly detection methods depending on the available knowledge of the dataset:

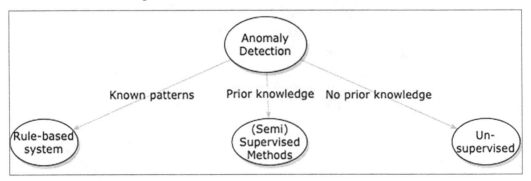

Figure 6.4: Anomaly detection methods depending on available knowledge

The earliest examples for anomaly detection consist of rule-based systems. This works when the patterns can be clearly defined. When we have an annotated set of anomalies, we can apply supervised or semi-supervised methods such as classifiers or regression models. The most common use case, however, is when anomalies are not annotated and we need unsupervised approaches to detect anomalous points or sections based on densities or distributions.

It's instructive to see what the big technology companies, Alphabet (Google), Amazon, Facebook, Apple, and Microsoft (GAFAM), do for anomaly detection. Let's go over each in turn and see how they handle anomaly detection.

Microsoft

In the paper *"Time-Series Anomaly Detection Service at Microsoft"* (Hansheng Ren and others, 2019) a time-series service is presented that's deployed for anomaly detection for production data at Microsoft. Its core is a **Spectral Residual** (**SR**) and Convolutional Neural Network (CNN) that's applied to unsupervised online anomaly detection in univariate time-series.

They borrowed the SR method from the concept of the saliency map in vision. Saliency maps highlight the points in images, which stand out to human observers. The algorithm performs the Fourier Transform of the data, then it applies the SR of the log amplitude of the transformed signal, and finally projects the spectral data back to the temporal domain with the Inverse Fourier Transform.

As an extension, they train a CNN based on artificial data using the SR method. They show benchmarks on publicly available data that support their claim that their method is the state of the art for anomaly detection.

They further claim that their detection accuracy (F1-score) is improved by more than 20% on Microsoft production data. You can find the basic implementation in the alibi-detect library (the *"Spectral Residual"* method).

Google

In their frequently asked questions on Google Analytics (`https://support.google.com/analytics/answer/7507748?hl=en`), Google refer to a Bayesian state space-time-series model (*"Predicting the Present with Bayesian Structural Time-Series"* by Steven L. Scott and Hal Varian, 2013) for change point and anomaly detection.

Google released an R package with more specific time-series functionality – CausalImpact. A paper describing the research behind the package was published in 2015 (*"Inferring causal impact using Bayesian structural time-series models"* by Kay H. Brodersen, Fabian Gallusser, Jim Koehler, Nicolas Remy, Steven L. Scott). CausalImpact estimates the causal effect of interventions based on a structural Bayesian time-series model. This has been ported to Python (the pycausalimpact library).We are going to experiment with causal impact analysis using Bayesian Structural Time-Series (BSTS) in *Chapter 9, Probabilistic Models for Time-Series*.

Amazon

Amazon, providing machine learning solutions at scale through their **Amazon Web Services (AWS)** platform, have anomaly detection as part of their resource and application monitoring solution, CloudWatch. It's unclear how their solution works, but Corey Quinn, an economist, theorized in a tweet that their solution is exponential smoothing. As part of that it is likely that they apply seasonal decomposition as a first step of their algorithm.

They also have a second service for anomaly detection: Amazon Lookout for Metrics. It's also unclear how this works under the hood. The service is geared toward monitoring business indicators, and – according to the documentation – is used internally within Amazon for large-scale monitoring. In the service, users can select fields from data sources with different breakdowns, for example, by selecting database columns `page_views` and `device_type`, users could look for abnormal changes in page views for every device type separately.

As for Amazon research in anomaly detection, they clinched the top three spots, out of 117 submissions in a challenge at the Workshop on the **Detection and Classification of Acoustic Scenes and Events (DCASE** 2020). This is a challenge comparable to time-series anomaly detection. They won the best paper award with *"Group Masked Autoencoder Based Density Estimator for Audio Anomaly Detection"* (Ritwik Giri and others, 2020).

Facebook

Facebook's Core Data Science team open-sourced their implementation for time-series forecasting and anomaly detection on GitHub. Their library is called Prophet. In their blog post announcing the library in 2017, they state that Prophet had been a key piece in Facebook's ability to create forecasts at scale and trusted as an important piece of information in decision-making.

The paper *"Forecasting at scale"* by Sean J Taylor and Benjamin Letham (2017) describes their setup at Facebook that includes an analyst in the loop and can automatically flag forecasts for manual review and adjustment. The anomaly detection builds on the uncertainty around the forecast from a Generalized Additive Model (GAM).

Prophet has been compared in benchmarks to other probabilistic and non-probabilistic models, and has rarely shown outstanding success. The Elo ratings at microprediction.com indicate that Prophet performs worse at univariate forecasts than exponential moving averages and many other standard methods.

Twitter

Twitter released an R package as well, called AnomalyDetection. Their method is based on the Generalized Extreme Studentized Deviate (ESD) test for detecting anomalies in univariate approximately normally distributed time-series. Their method was published in 2017 (*"Automatic Anomaly Detection in the Cloud Via Statistical Learning"*, Jordan Hochenbaum, Owen Vallis, Arun Kejariwal).

For their adaption of the ESD test, the Seasonal Hybrid ESD, they included a Seasonal-Trend decomposition using LOESS (STL) before applying a threshold on the z-score (as mentioned above) or – for datasets with a high number of anomalies – thresholding based on the median and MAD. The Twitter model has been ported to Python (the sesd library).

Implementations

We'll end with an overview of readily available implementations for anomaly detection in Python. Lots of implementations are available. Their use cases are very similar, however, the implementations and the user bases are widely different.

Here's a list ordered by the number of stars on GitHub (as per May 2021):

Library	Implementations	Maintainer	Stars
Prophet	Uncertainty interval around the estimated trend component from the forecast	Facebook Core Research	12.7k
PyOD	30 detection algorithms for multivariate time-series — from classical LOF (SIGMOD 2000) to COPOD (ICDM 2020)	Yue Zhao and others	4.5k
alibi-detect	Many anomaly detection algorithms — specific to time-series there are Likelihood Ratios, Prophet, Spectral Residual, Seq2Seq, Model distillation	Seldon Technologies Ltd	683
Scikit-Lego	Reconstruction through PCA/UMAP	Vincent D. Warmerdam and others	499
Luminaire	Luminaire Window Density Model	Zillow	371
Donut	Variational Auto-Encoder for Seasonal KPIs	Tsinghua Netman Lab	327
rrcf	Robust Random Cut Forest algorithm for anomaly detection on streams	Real-time water systems lab, University of Michigan	302
banpei	Hotelling's theory	Hirofumi Tsuruta	245
STUMPY	Matrix Profile algorithms for uni- and multivariate time-series such as STUMP, FLUSS, and FLOSS (also compare matrixprofile-ts)	TD Ameritrade	169
PySAD	More than a dozen algorithms for streaming outlier anomaly detection	Selim Yilmaz, Selim and Suleyman Kozat	98

Figure 6.5: Anomaly detection methods in Python

Each of these methods has their own background and formal underpinning; however, it's out of scope in this chapter to describe all of them.

This chart shows the star history (from star-history.t9t.io) of the top three repositories:

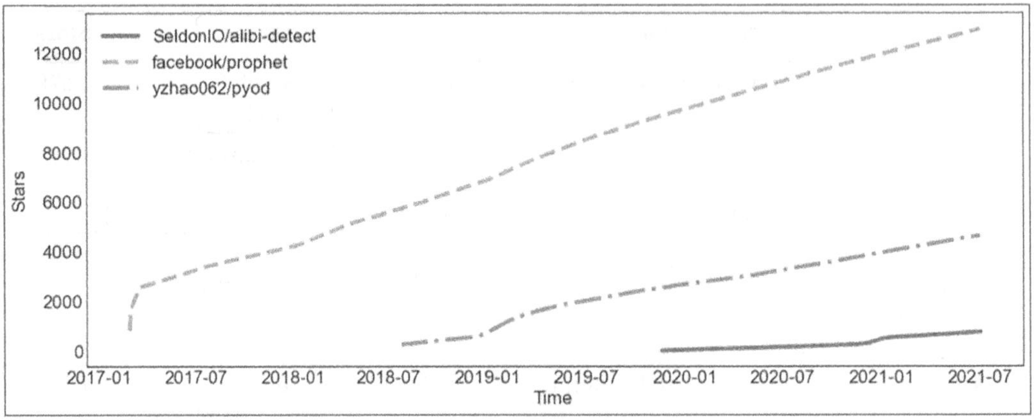

Figure 6.6: Star history of Prophet, PyOD, and alibi-detect

Both Prophet and PyOD have been seeing a continuous rise in popularity (GitHub stars).

Many deep learning algorithms have been applied more recently to anomaly detection, both with univariate and multivariate time-series.

What's particularly interesting about deep learning models is that the applications can be much broader: anomaly detection in video surveillance in closed-circuit television. We'll go more into detail of deep learning architectures in *Chapter 10, Deep Learning for Time-Series*.

Change point detection

A common problem with time-series is changes in the behavior of the observed system. Generally speaking, a change point signals an abrupt and significant transition between states in the process generating the series. For example, the trend can suddenly change, and a change point can signal where the trend of the series changes. This is well known under the guise of technical chart pattern analysis in trading.

This list captures some applications for **Change point detection (CPD)**:

- Speech recognition: Detection of word and sentence boundaries
- Image analysis: Surveillance on video footage of closed-circuit television
- Fitness: Segmenting human activities over time based on data from motion sensors from smart devices such as watches or phones

- Finance: Identifying changes to trend patterns that could indicate changes from bear to bull markets, or the other way around

As an example for the importance of CPD, consider the stock market. Time-series data that describes the evolution of a market, such as stock prices, follows trends – it either rises, falls, or doesn't change significantly (stagnation).

When a stock rises, the investor wants to buy the stock. Otherwise, when the stock is falling, the investors doesn't want to keep the stock, but rather to sell it. Not changing the position will cause a loss of book value – in the best case, this will cause a problem with liquidity.

For investors, it is therefore key to know, when the market changes from rising to falling or the other way around. Recognizing these changes can make the difference between winning or losing.

In forecasting, special events like Black Friday, Christmas, an election, a press release, or changes in regulation can cause short-term (perhaps then classed as an anomaly) or long-term change to the trend or to the level of the series. This will inevitably lead to strange predictions from traditional models.

A particularly interesting challenge with CPD algorithms is detecting these inflection points in real time. This means detecting a change point as soon as it arrives (or, at the very least, before the next change point occurs).

We can distinguish online and offline methods for CPD, where online refers to processing on the fly, dealing with each new data point as it becomes available. On the other hand, offline algorithms can work on the whole time-series at once. We'll deal more with online processing in *Chapter 8, Online Learning for Time-Series*.

CPD is related to segmentation, edge detection, event detection, and anomaly detection, and similar techniques can be applied to all these applications. CPD can be viewed as very much like anomaly detection, since one way to identify change points is by anomaly scores from an anomaly detection algorithm.

From this perspective, change points are identical to highly anomalous points, and anything above a certain threshold corresponds to a change. In the same way as anomaly detection, CPD can be defined as the problem of hypothesis testing between two alternatives, the null hypothesis being *"no change occurs,"* and the alternative hypothesis of *"a change occurs."*

CPD algorithms are composed of three components: cost functions, search methods, and constraints. We'll go through these in turn. Cost functions are distance functions that can be applied to a subsection of the time-series (multivariate or univariate).

An example for a cost function is **least absolute deviation** (LAD), which is an estimator of a shift in the central point (mean, median, and mode) of a distribution defined as follows:

$$c(x_I) = \sum_{t \in I} |x_t - \hat{x}|$$

In this definition l is an index to a subsection in the time-series x, and $\hat{x}$ is the central point of x.

The search function then iterates over the time-series to detect change points. This can be done approximately, such as in window-based detection, bottom-up methods, or binary segmentation, or it can be exhaustive as in the case of dynamic programming or **Pruned Exact Linear Time** (**Pelt**).

Pelt (Gachomo Dorcas Wambui and others, 2015) relies on pruning heuristics, and has a computational cost that is linear to the number of points of the time-series, $O(T)$. Dynamic programming methods have a much higher computational cost of $O(nT^2)$, where n is the maximum number of expected change points.

Finally, the constraint can come into play as a penalty in the search algorithm. This penalty term can encode a cost budget or knowledge of the number of change points that we would expect to find.

It is notoriously difficult to evaluate the performance of CPD algorithms, because of the lack of benchmark datasets. Only very recently (2020) Gerrit van den Burg and Christopher Williams from the Alan Turing Institute and the University of Edinburgh published a benchmark consisting of 37 time-series from sources such as the World Bank, EuroStat, U.S. Census Bureau, GapMinder, and Wikipedia. Their benchmark is available on GitHub, and they mention change point annotations for datasets centered around the financial crisis of 2007-08, legislation on seat belts in the U.K., the Montreal Protocol regulating chlorofluorocarbon emissions, or the regulation of automated phone calls in the U.S.

In the same paper ("*An Evaluation of Change Point Detection Algorithm*"), the authors evaluated a whole range of methods for CPD. They note that their "zero" baseline method, which assumes no change points all, outperforms many of the other methods, according to F1-measure and a cluster overlap measure based on the Jaccard index. This is because of the small proportion of change points in the dataset, and the high number of false positives the methods return. They concluded that binary segmentation and Bayesian online CPD are among the best methods across the time-series.

Binary segmentation ("*On Tests for Detecting Change in Mean*" by Ashish Sen and Muni S. Srivastava, 1975) falls into the category of window-based CPD. Binary segmentation is a greedy algorithm that minimizes the sum of costs the most as defined like this:

$$\hat{t} := argmin_{1 \leq t < T-1} c(x_{0 \cdots t}) + c(x_{t \cdots T})$$

$\hat{t}$ is the found change point and $c()$ is a cost function similar to LAD, which we saw earlier in this section. The general idea is that when two subsequences are highly dissimilar, this indicates a change point.

Binary segmentation is sequential in the sense that the change point is detected first on the complete time-series, then again on the two sub-sequences before and after the change point. This explains its low complexity of $O(T \log T)$, where T is the length of the time-series. This computational cost makes it scalable to larger datasets.

This table presents an overview of methods for CPD:

Library	Implementations	Maintainer
Greykite	CPD via adaptive lasso	LinkedIn
ruptures	Offline CPD: binary segmentation, dynamic programming, Pelt, window-based	Charles Truong
Bayesian Changepoint Detection	Bayesian CPD	Johannes Kulick
banpei	Singular spectrum transformation	Hirofumi Tsuruta
changepy	Pelt algorithm	Rui Gil
onlineRPCA	Online Moving Window Robust Principal Component Analysis	Wei Xiao

Figure 6.7: CPD methods in Python

We've omitted Facebook's Prophet library since it's not a dedicated CPD package.

The chart below illustrates the popularity of CPD methods over time.

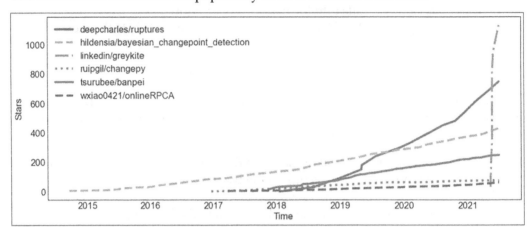

Figure 6.8: Star history of CPD methods

LinkedIn's Greykite has been seeing a meteoric rise in GitHub stars since its release. Also ruptures has seen a huge increase in popularity.

Clustering

Cluster analysis or clustering is the process of finding meaningful groups (clusters) of points or objects in a dataset based on their similarity. As the result of this unsupervised data mining technique, we want points in each cluster to be similar to each other, while being different to points in other clusters.

Clustering of time-series is challenging because each data point is a period of time (an ordered sequence). It has found application in diverse areas to discover patterns that empower time-series analysis, extracting insights from complex datasets.

We are not going to get into details on time-series clustering, but the following table gives an overview of Python libraries for time-series clustering:

Library	Implementations	Maintainer	Stars
tslearn	Time-Series K-Means, K-Shape clustering, KernelKMeans	Romain Tavenard	1.7k
river	DBStream, Time-Series K-Means, CluStream, DenStream, STREAMKMeans	Albert Bifet and others	1.7k

Figure 6.9: Clustering methods for time-series in Python

You can see the GitHub stars for the top implementation over history here:

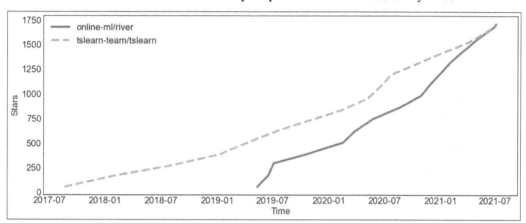

Figure 6.10: Star history of tslearn and river

Both libraries are going strong. We'll revisit river in *Chapter 8, Online Learning for Time-Series*.

Python practice

Let's do first an example of anomaly detection, then another for CPD. Let's first look at the needed libraries in the next section.

Requirements

In this chapter, we'll use several libraries, which we can quickly install from the terminal (or similarly from the anaconda navigator):

```
pip install ruptures alibi_detect
```

We'll execute the commands from the Python (or IPython) terminal, but equally we could execute them from a Jupyter notebook (or a different environment).

We should be ready now to get into the woods with implementing unsupervised time-series algorithms in Python.

Anomaly detection

alibi-detect comes with several benchmark datasets for time-series anomaly detection:

- fetch_ecg — ECG dataset from the BIDMC Congestive Heart Failure Database
- fetch_nab — Numenta Anomaly Benchmark
- fetch_kdd — KDD Cup '99 dataset of computer network intrusions

The last of these is loaded through scikit-learn.

Let's load the time-series of computer network intrusions (KDD99):

```
from alibi_detect.datasets import fetch_kdd
intrusions = fetch_kdd()
```

intrusions is a dictionary, where the data key returns a matrix of 494021x18. The 18 dimensions of the time-series are the continuous features of the dataset, mostly error rates and counts:

```
intrusions['feature_names']
['srv_count',
 'serror_rate',
 'srv_serror_rate',
 'rerror_rate',
 'srv_rerror_rate',
 'same_srv_rate',
 'diff_srv_rate',
 'srv_diff_host_rate',
 'dst_host_count',
 'dst_host_srv_count',
 'dst_host_same_srv_rate',
 'dst_host_diff_srv_rate',
 'dst_host_same_src_port_rate',
 'dst_host_srv_diff_host_rate',
 'dst_host_serror_rate',
 'dst_host_srv_serror_rate',
 'dst_host_rerror_rate',
 'dst_host_srv_rerror_rate']
```

Another key, target contains the annotations of anomalies.

Since we have the annotations ready we could train a classifier, however, we'll stick to unsupervised methods. Further, since the Spectral Method that we'll use is for univariate data and we'll only take a single dimension out of our multivariate dataset, we'll completely ignore the annotations.

Here's a quick plot of our time-series (we'll choose – arbitrarily – the first dimension of our dataset):

```
import pandas as pd
pd.Series(intrusions['data'][:, 0]).plot()
```

This is the plot:

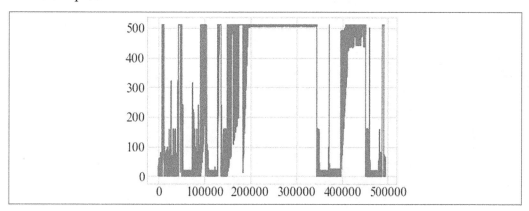

Figure 6.11: Time-series chart

We'll load and run the SpectralResidual model that implements the method proposed by Microsoft:

```
from alibi_detect.od import SpectralResidual
od = SpectralResidual(
    threshold=1.,
    window_amp=20,
    window_local=20,
    n_est_points=10,
    n_grad_points=5
)
```

We can then get the anomaly scores for each point in our time-series:

```
scores = od.score(intrusions['data'][:, 0])
```

Let's plot the scores imposed on top of our time-series!

```
import matplotlib

ax = pd.Series(intrusions['data'][:, 0], name='data').
plot(legend=False, figsize=(12, 6))
ax2 = ax.twinx()
ax = pd.Series(scores, name='scores').plot(ax=ax2, legend=False,
color="r", marker=matplotlib.markers.CARETDOWNBASE)
ax.figure.legend(bbox_to_anchor=(1, 1), loc='upper left');
```

We are using a dual y-axis for plotting the scores and the data within the same plot. Here it is:

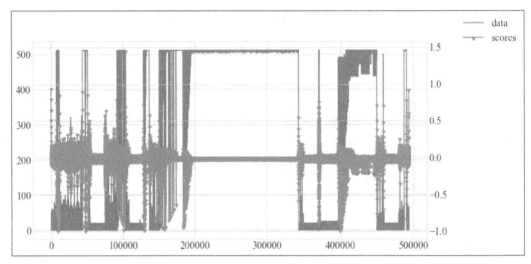

Figure 6.12: Time-series with anomalies

Some points are not recognized as outliers since the periodic nature of the signal is removed by the Fourier filter.

Change point detection

We'll first create a synthetic multivariate time-series with the ruptures library. We'll set the number of dimensions to 3 and the length of the time-series to 500, and our time-series will have 3 change points and a Gaussian noise of standard deviation 5.0 will be over imposed:

```
import numpy as np
import matplotlib.pylab as plt
import ruptures as rpt
```

```
signal, bkps = rpt.pw_constant(
  n_samples=500, n_features=3, n_bkps=3,
  noise_std=5.0, delta=(1, 20)
)
```

Signal is a NumPy array of 500x3. bkps is the array of the change points (123, 251, and 378).

We can plot this time-series with a utility function that highlights the subsections separated by changepoints:

```
rpt.display(signal, bkps)
```

Here's the plot of our time-series with three change points:

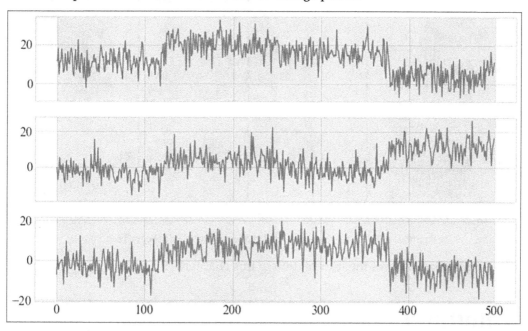

Figure 6.13: Time-series with change points

We can apply Binary Segmentation to this time-series. ruptures follows the scikit-learn conventions, so if you have used scikit-learn before, the usage should be very intuitive:

```
algo = rpt.Binseg(model="l1").fit(signal)
my_bkps = algo.predict(n_bkps=3)
```

We have several options for the Binary Segmentation constraint – we have the choice between `l1`, `l2`, `rbf`, `linear`, `normal`, and `ar`.

We can plot the predictions of the Binary Segmentation with another utility function:

```
rpt.show.display(signal, bkps, my_bkps, figsize=(10, 6))
```

Here's the plot of our change point predictions from the Binary Segmentation model:

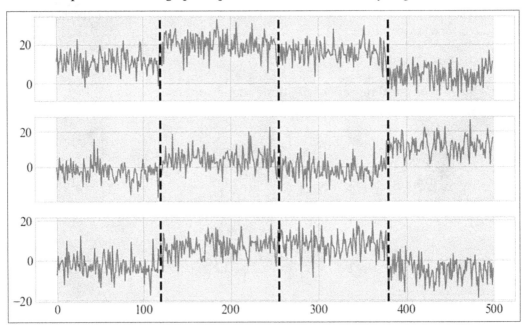

Figure 6.14: Time-series with detected change points (Binary Segmentation)

Let's summarize some of the information in this chapter!

Summary

In this chapter, we have concentrated on two aspects of unsupervised methods for time-series:

- Anomaly detection
- Change point detection

The essence of anomaly detection (also: outlier detection) is to identify sequences that are notably different from the rest of the series. We've investigated different anomaly detection methods, and how several major companies are dealing with it at scale.

When working with time-series, it's important to be aware of changes in the data over time that makes models useless (model staleness). This is called change point detection and drift detection.

We've looked at change point detection in this chapter. In *Chapter 8, Online Learning for Time-Series*, we'll look at drift detection in more detail.

7

Machine Learning Models for Time-Series

Machine learning has come a long way in recent years, and this is reflected in the methods available to time-series predictions. We've introduced a few state-of-the-art machine learning methods for time-series in *Chapter 4, Introduction to Machine Learning for Time-Series*. In the current chapter, we'll introduce several more machine learning methods.

We'll go through methods that are commonly used as baseline methods, or that stand out in terms of either performance, ease of use, or their applicability. I'll introduce k-nearest neighbors with dynamic time warping and gradient boosting for time-series as a baseline and we'll go over other methods, such as Silverkite and gradient boosting. Finally, we'll go through an applied exercise with some of these methods.

We're going to cover the following topics:

- More machine learning methods for time-series
- K-nearest neighbors with dynamic time warping
- Silverkite
- Gradient boosting
- Python exercise

If you are looking for a discussion of state-of-the-art machine learning algorithms, please refer to *Chapter 4, Introduction to Machine Learning for Time-Series*. The discussion of algorithms will assume some of the information of that chapter. The algorithms that we'll cover in the next sections are all highly competitive for forecasting and prediction tasks.

We'll discuss algorithms here in more detail.

More machine learning methods for time-series

The algorithms that we'll cover in this section are all highly competitive for forecasting and prediction tasks. If you are looking for a discussion of state-of-the-art machine learning algorithms, please refer to *Chapter 4, Introduction to Machine Learning for Time-Series*.

In the aforementioned chapter, we've briefly discussed a few of these algorithms, but we'll discuss them here in more detail and we will also introduce other algorithms that we haven't discussed before, such as Silverkite, gradient boosting, and k-nearest neighbors.

We'll dedicate a separate practice section to a library that was released in 2021, which is facebook's Kats. Kats provides many advanced features, including hyperparameter tuning and ensemble learning. On top of these features, they implement feature extraction based on the TSFresh library and include many models, including Prophet, SARIMA, and others. They claim that their hyperparameter tuning for time-series is about 6-20 times faster in benchmarks compared with other hyperparameter tuning algorithms.

This graph provides an overview of the popularity of selected time-series machine learning libraries:

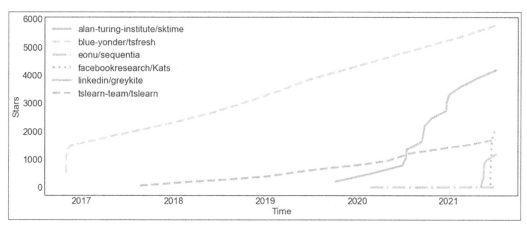

Figure 7.1: Popularity of time-series machine learning libraries

As of mid-2021, Kats and GreyKite have been released very recently, and although they have been garnering stars on GitHub, they haven't accumulated enough to rival TSFresh's popularity. I've included TSFresh even though it is a library for feature generation, and not prediction. I found it interesting to see how important it is in relation to other libraries that we use in this chapter. After TSFresh, SKTime is second, and it has been attracting a lot of stars over a relatively short time period.

We'll use a few of these libraries in the practical examples in this chapter.

Another important issue is validation, and it's worth covering this separately.

Validation

We've discussed validation before in *Chapter 4, Introduction to Machine Learning for Time-Series*. Often, in machine learning tasks, we use k-fold cross-validation, where splits of the data are performed pseudo-randomly, so the training and the test/validation datasets can come from any part of the data as long as it hasn't been used for training (**out-of-sample data**).

With time-series data, this way of validation can lead to an overconfidence in the model's performance because, realistically, time-series tend to change over time according to trend, seasonality, and changes to the time-series characteristics.

Therefore, with time-series, validation is often performed in a so-called **walk-forward validation**. This means that we train the model on past data, and we'll test it on the newest slice of data. This will take out the optimistic bias and give us a more realistic estimate of performance once the model is deployed.

In terms of training, validation, and test datasets, this means that we'll adjust model parameters entirely on training and validation datasets, and we'll benchmark our test based on a set of data that's more advanced in time, as illustrated in the following diagram (source: Greykite library's GitHub repository):

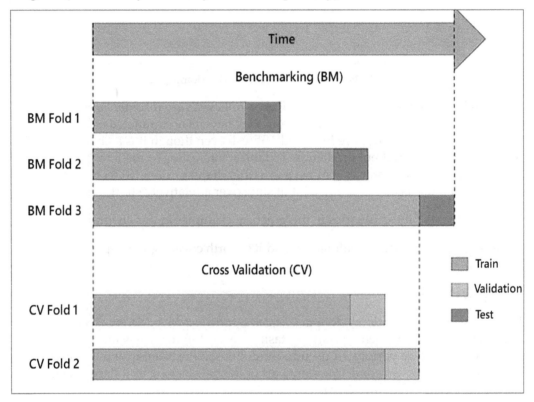

Figure 7.2: Walk-forward validation

In walk-forward validation, we train on an initial segment of the data and then test on a period after the training set. Next, we roll forward and repeat the process. This way, we have multiple out-of-sample periods and can combine the results over these periods. With walk-forward, we are less likely to suffer from overfitting.

K-nearest neighbors with dynamic time warping

K-nearest neighbors is a well-known machine learning method (sometimes also going under the guise of case-based reasoning). In kNN, we can use a distance measure to find similar data points. We can then take the known labels of these nearest neighbors as the output and integrate them in some way using a function.

Figure 7.3 illustrates the basic idea of kNN for classification (source – WikiMedia Commons: `https://commons.wikimedia.org/wiki/File:KnnClassification.svg`):

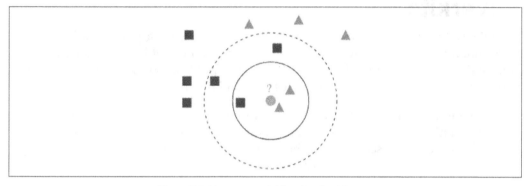

Figure 7.3: K-nearest neighbor for classification

We know a few data points already. In the preceding illustration, these points are indicated as squares and triangles, and they represent data points of two different classes, respectively. Given a new data point, indicated by a circle, we find the closest known data points to it. In this example, we find that the new point is similar to triangles, so we might assume that the new point is of the triangle class as well.

While this method is conceptually very simple, it often serves as a strong baseline method, or is sometimes even competitive with more sophisticated machine learning algorithms, even when we only compare the closest neighbor *(k=1)*.

The important hyperparameters in this algorithm are:

- The number of neighbors (k) you want to base your output on
- The integration function (for example, the average or the most frequent value)
- The distance function to use to find the nearest data points

We talked about dynamic time warping in *Chapter 4, Introduction to Machine Learning for Time-Series*, as a measure that can be used to compare the similarity (or, equivalently, the distance) between two time-series. These sequences can even be of different lengths. Dynamic time warping has proven itself to be an exceptionally strong distance measure for time-series.

We can use kNN in combination with dynamic time warping as a distance measure to find similar time-series, and this method has proven itself hard to beat, although the state of the art has since surpassed it.

Silverkite

The Silverkite algorithm ships together with the Greykite library released by LinkedIn. It was explicitly designed with the goals in mind of being fast, accurate, and intuitive. The algorithm is described in a 2021 publication ("*A flexible forecasting model for production systems*", by Reza Hosseini and others).

According to LinkedIn, it can handle different kinds of trends and seasonalities such as hourly, daily, weekly, repeated events, and holidays, and short-range effects. Within LinkedIn, it is used for both short-term, for example, a 1-day head, and long-term forecast horizons, such as 1 year ahead.

Use cases within LinkedIn include optimizing budget decisions, setting business metric targets, and providing sufficient infrastructure to handle peak traffic. Furthermore, a use case has been to model recoveries from the COVID-19 pandemic.

The time-series is modeled as an additive composite of trends, change points, and seasonality, where seasonality includes holiday/event effects. The trend is then modeled as follows:

$$trend(t) = \alpha_0 f(t) + \sum_{i=1}^{K} \alpha_i \mathbf{1}_{t>t_i}(f(t) - f(t_i))$$

where K is the number of change points, and t_i is the time index for the i-th change point. Therefore, $\mathbf{1}_{t>t_i}$ is an indicator function for the i-th change point. The function f(t) can be linear, square root, quadratic, any combination, or completely custom.

Silverkite also constructs indicator variables for holidays. Holidays can be specified by name or by country, or can even be completely custom.

Change points can be specified manually or candidates can be automatically detected with a regression model and subsequently selected using the Adaptive Lasso algorithm (Hui Zhou, 2006).

In addition to trend, seasonality, and holiday, Silverkite includes an autoregressive term that is calculated based on windowed averages rather than taking lags independently (*"Selecting a binary Markov model for a precipitation process"*, by Reza Hosseini and others, 2011).

This autoregressive term is specified using the Pasty library, using a formula mini-language, in a form like this string:

```
y ~ a + a:b + np.log(x)
```

In this formula, y on the left-hand side is defined as the sum of three terms, a, a:b, and np.log(x). The term a:b is an interaction between two factors, a and b. The model template itself in Pasty is highly customizable, so this interface provides a high degree of flexibility.

Finally, Silverkite comes with several model types, such as ridge regression, elastic net, and boosted trees, with supported loss functions, MSE, and quantile loss for robust regression.

According to a LinkedIn benchmark on several datasets, Silverkite outperforms both auto-Arima (the pmdarima library) and Prophet in terms of prediction error. Yet, Silverkite was about four times as fast as Prophet, which we'll introduce in *Chapter 9, Probabilistic Models*.

Gradient boosting

XGBoost (short for **eXtreme Gradient Boosting**) is an efficient implementation of gradient boosting (Jerome Friedman, *"Greedy function approximation: a gradient boosting machine"*, 2001) for classification and regression problems. Gradient boosting is also known as **Gradient Boosting Machine (GBM)** or **Gradient Boosted Regression Tree (GBRT)**. A special case is LambdaMART for ranking applications. Apart from XGBoost; other implementations are Microsoft's Light Gradient Boosting Machine (LightGBM), and Yandex's Catboost.

Gradient Boosted Trees is an ensemble of trees. This is similar to Bagging algorithms such as Random Forest; however, since this is a boosting algorithm, each tree is computed to incrementally reduce the error. With each new iteration a tree is greedily chosen and its prediction is added to the previous predictions based on a weight term. There is also a regularization term that penalizes complexity and reduces overfitting, similar to the Regularized Greedy Forest (RGF).

The **XGBoost** algorithm was published in 2016 by Tianqi Chen and Carlos Guestrin ("*XGBoost: A Scalable Tree Boosting System*") and pushed the envelope on many classification and regression benchmarks. It was used in many winning solutions to Kaggle problems. In fact, in 2015, of the 29 challenge-winning solutions, 17 solutions used XGBoost.

It was designed to be highly scalable and features extensions of the gradient boosting algorithm for weighted quantiles, along with improvements for scalability and parallelization based on smarter caching patterns, sharding, and the handling of sparsity.

As a special case of regression, XGBoost can be used for forecasting. In this scenario, the model is trained based on past values to predict future values, and this can be applied to univariate as well as multivariate time-series.

Python exercise

Let's put into practice what we've learned in this chapter so far.

As for requirements, in this chapter, we'll be installing requirements for each section separately. The installation can be performed from the terminal, the notebook, or from the anaconda navigator.

In a few of the following sections, we'll demonstrate classification in a forecast, so some of these approaches will not be comparable. The reader is invited to do forecasts and classification using each approach and then compare results.

As a note of caution, both Kats and Greykite (at the time of writing) are very new libraries, so there might still be frequent changes to dependencies. They might pin your NumPy version or other commonly used libraries. Therefore, I'd recommend you install them in virtual environments separately for each section.

We'll go through this setup in the next section.

Virtual environments

In a Python virtual environment, all libraries, binaries, and scripts installed into it are isolated from those installed in other virtual environments and from those installed in the system. This means that we can have different libraries, such as Kats and Greykite, installed without having to bother with compatibility issues between them or with other libraries installed on our computer.

Let's go through a quick tutorial introduction to using virtual environments with Jupyter notebooks using anaconda (similarly, you can use tools such as virtualenv or pipenv).

In *Chapter 1*, *Introduction to Time-Series with Python*, we went through the installation of Anaconda, so we'll skip the installation. Please refer to that chapter or go to conda. io for instructions.

To create a virtual environment, you have to specify a name:

```
conda create --name myenv
```

This will create an eponymous directory (myenv), where all libraries and scripts will be installed.

If we want to use this environment, we have to activate it first, which means that we set the PATH variable to include our newly created directory:

```
conda activate myenv
```

We can now use tools such as pip, which will default to the one bundled with conda, or the conda command directly to install libraries.

We can install Jupyter or Jupyter labs into our environment and then start it. This means that our Jupyter environment will include all dependencies as we've installed them in isolation.

Let's start with a kNN algorithm with dynamic time warping. As I've mentioned, this algorithm often serves as a decent baseline for comparisons.

K-nearest neighbors with dynamic time warping in Python

In this section, we'll classify failures from force and torque measurements of a robot over time.

We'll use a very simple classifier, kNN, and perhaps we should give a heads-up that this method involves taking point-wise distances, which can often be a bottleneck for computations.

In this section, we'll combine TSFresh's feature extraction in a pipeline with a kNN algorithm. The time-series pipeline can really help make things easy, as you'll find when reading the code snippets.

Let's install tsfresh and tslearn:

```
pip install tsfresh tslearn
```

We'll use the kNN classifier in tslearn. We could even have used the kNN classifier in scikit-learn, which allows a custom metric to be specified.

In the example, we will download a dataset of robotic execution failures from the UCI machine learning repository and store it locally. This dataset contains force and torque measurements on a robot after failure detection. For each sample, the task is to classify whether the robot will report a failure:

```
from tsfresh.examples import load_robot_execution_failures
from tsfresh.examples.robot_execution_failures import download_robot_
execution_failures

download_robot_execution_failures()
df_ts, y = load_robot_execution_failures()
```

The columns include the time and six time-series with signals from the sensors, F_x, F_y, F_z, T_x, T_y, and T_z. The target variable, y, which can take the values True or False, indicates if there was a failure.

It's always important to check the frequency of the two classes:

```
print(f"{y.mean():.2f}")
```

The mean of y is 0.24.

We can then extract time-series features using TSFresh, as discussed in *Chapter 3, Preprocessing Time-Series*. We can impute missing values and select features based on relevance to the target. In TSFresh, the p-value from a statistical test is used to calculate the feature significance:

```
from tsfresh import extract_features
from tsfresh import select_features
from tsfresh.utilities.dataframe_functions import impute

extracted_features = impute(extract_features(df_ts, column_id="id",
column_sort="time"))
features_filtered = select_features(extracted_features, y)
```

We can continue working with the features_filtered DataFrame, which contains our features – sensor signals from before and TSFresh features.

Let's find some good values for the number of neighbors by doing a grid search:

```
from sklearn.model_selection import TimeSeriesSplit, GridSearchCV
from tslearn.neighbors import KNeighborsTimeSeriesClassifier

knn = KNeighborsTimeSeriesClassifier()
param_search = {
    'metric' : ['dtw'],
    'n_neighbors': [1, 2, 3]
}
tscv = TimeSeriesSplit(n_splits=2)

gsearch = GridSearchCV(
    estimator=knn,
    cv=tscv,
    param_grid=param_search
)
gsearch.fit(
    features_filtered,
    y
)
```

We are using scikit-learn's `TimeSeriesSplit` to split the time-series. This is for the GridSearch.

Alternatively, we could have just split based on an index.

There are many parameters we could have tried, especially for the distance metric in the kNN classifier. If you want to have a play with this, please see `TSLEARN_VALID_METRICS` for a complete list of metrics supported by tslearn.

Let's do a few forecasts of COVID cases. In the next section, we'll start with the Silverkite algorithm. Silverkite comes with the Greykite library released by LinkedIn in 2021.

Silverkite

At the time of writing, Greykite is in version 0.1.1 – it's not fully stable yet. Its dependencies might conflict with newer versions of commonly used libraries, including Jupyter Notebooks. Do not worry though if you install the library in your virtual environment or on Google Colab.

Just go ahead and install the library with all its dependencies:

```
pip install greykite
```

Now that Greykite is installed, we can use it.

We'll load up the COVID cases from the *Our World in Data* dataset, probably one of the best sources of available COVID data:

```
import pandas as pd

owid_covid = pd.read_csv("https://covid.ourworldindata.org/data/owid-covid-data.csv")
owid_covid["date"] = pd.to_datetime(owid_covid["date"])
df = owid_covid[owid_covid.location == "France"].set_index("date",
drop=True).resample('D').interpolate(method='linear')
```

We are concentrating on cases in France.

We start by setting up the Greykite metadata parameters. We'll then pass this object into the forecaster configuration:

```
from greykite.framework.templates.autogen.forecast_config import (
    ForecastConfig, MetadataParam
)
metadata = MetadataParam(
    time_col="date",
    value_col="new_cases",
    freq="D"
)
```

Our time column is date and our value column is new_cases.

We'll now create the forecaster object, which creates forecasts and stores the result:

```
import warnings
from greykite.framework.templates.forecaster import Forecaster
from greykite.framework.templates.model_templates import
ModelTemplateEnum

forecaster = Forecaster()
    warnings.filterwarnings("ignore", category=UserWarning)
    result = forecaster.run_forecast_config(
        df=yahoo_df,
```

```
config=ForecastConfig(
    model_template=ModelTemplateEnum.SILVERKITE_DAILY_90.name,
    forecast_horizon=90,
    coverage=0.95,
    metadata_param=metadata
)
)
```

The forecast horizon is 90 days; we will forecast 90 days ahead. Our prediction interval is 95%. Both Silverkite and Prophet support quantifying uncertainty by means of prediction intervals. A coverage of 95% means that 95% of actuals should fall within the prediction interval. In Greykite, the `_components.uncertainty` model provides additional configuration options about uncertainty.

I've added a line to ignore warnings of the `UserWarning` type during training since otherwise, there are about 500 lines of warnings about 0s in the target column.

Let's plot the original time-series from the result object. We can overlay our forecasts:

```
forecast = result.forecast
forecast.plot().show(renderer="colab")
```

Please leave out the `renderer` argument if you are not on Google Colab!

We get the following plot:

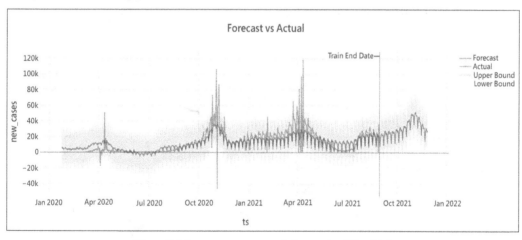

Figure 7.4: Forecast versus actual time-series (Silverkite)

The forecasts are in the `df` attribute of the `forecast` object:

```
forecast.df.head().round(2)
```

These are the upper and lower confidence intervals of the forecasts:

	date	actual	forecast	forecast_lower	forecast_upper
0	2020-01-24	2.0	6887.23	-10392.47	24166.93
1	2020-01-25	1.0	4797.20	-14791.35	24385.76
2	2020-01-26	0.0	6106.02	-20762.71	32974.75
3	2020-01-27	0.0	3693.54	-17082.31	24469.39
4	2020-01-28	1.0	4264.10	-11711.46	20239.65

Figure 7.5: Table of forecast versus actual time-series (Silverkite)

We might want to get some performance metrics for our model. We can get the performance of the historical forecast on the holdout test set like this:

```
from collections import defaultdict
backtest = result.backtest

backtest_eval = defaultdict(list)
for metric, value in backtest.train_evaluation.items():
    backtest_eval[metric].append(value)
    backtest_eval[metric].append(backtest.test_evaluation[metric])
metrics = pd.DataFrame(backtest_eval, index=["train", "test"]).T
metrics.head()
```

Our performance metrics look like this:

	train	test
CORR	0.798556	0.410079
R2	0.625995	-0.706472
MSE	1.00066e+08	1.85563e+08
RMSE	10003.3	13622.2
MAE	4833.13	11008.9

Figure 7.6: Performance metrics on the hold-out data (Silverkite)

I've truncated the metrics to the first five.

We can apply our model conveniently to new data like this:

```
model = result.model
future_df = result.timeseries.make_future_dataframe(
    periods=4,
    include_history=False
)
model.predict(future_df)
```

The predictions look like this:

	ts	forecast	forecast_lower	forecast_upper	y_quantile_summary	err_std
0	2021-08-30	8149.801170	-12626.050813	28925.653153	(-12626.050812949516, 28925.65315263861)	10600.119261
1	2021-08-31	18795.031174	2819.476908	34770.585441	(2819.4769075703443, 34770.58544098518)	8150.942769
2	2021-09-01	22903.216512	-2036.397958	47842.830983	(-2036.3979580955383, 47842.83098281716)	12724.526913
3	2021-09-02	24881.782865	4326.654006	45436.911725	(4326.654005899352, 45436.911724790116)	10487.503353

Figure 7.7: Forecast DataFrame (Silverkite)

Please note that your result might vary.

We can use other forecaster models by changing the `model_template` argument in the run configuration of the forecaster. For instance, we could set it to `ModelTemplateEnum.PROPHET.name` in order to take Facebook's Prophet model.

This concludes our tour of Silverkite. Next, we will forecast by applying a supervised regression method with XGBoost. Let's do some gradient boosting!

Gradient boosting

We can use supervised machine learning for time-series forecasting as well. For this, we can use the dates and the previous values to predict the future.

First, we need to install XGBoost:

```
pip install xgboost
```

We'll use the Yahoo daily closing data in this example, as in other practice sections of this chapter.

Let's go through the preparation and modeling step by step.

We first need to featurize the data. Here, we'll do this by extracting date features, but please see the section on kNNs, where TSFresh's feature extraction is used instead. You might want to change this example by combining the two feature extraction strategies or by relying on TSFresh entirely.

We will reload the new COVID cases from the *Our World in Data* dataset as before:

```python
import pandas as pd

owid_covid = pd.read_csv("https://covid.ourworldindata.org/data/owid-covid-data.csv")
owid_covid["date"] = pd.to_datetime(owid_covid["date"])
df = owid_covid[owid_covid.location == "France"].set_index("date",
drop=True).resample('D').interpolate(method='linear').reset_index()
```

For feature extraction, transformers are handy. A transformer is basically a class with `fit()` and `transform()` methods that can make the transformer adapt to a dataset and transform the data accordingly. Here's the code for the `DateFeatures` transformer that annotates a dataset according to a date:

```python
from sklearn.base import TransformerMixin, BaseEstimator

class DateFeatures(TransformerMixin, BaseEstimator):
    features = [
        "hour",
        "year",
        "day",
        "weekday",
        "month",
        "quarter",
    ]

    def __init__(self):
        super().__init__()

    def transform(self, df: pd.DataFrame):
        Xt = []
        for col in df.columns:
            for feature in self.features:
                date_feature = getattr(
                    getattr(
                        df[col], "dt"
                    ), feature
```

```
            )
            date_feature.name = f"{col}_{feature}"
            Xt.append(date_feature)

        df2 = pd.concat(Xt, axis=1)
        return df2

    def fit(self, df: pd.DataFrame, y=None, **fit_params):
        return self
```

This transformer is relatively simple in that it extracts a range of features for a date column such as hours, years, days, weekday, months, week of year, and quarter. These features can potentially be very powerful for describing or annotating the time-series data in a machine learning context.

You can find the complete code for this example on GitHub. I am providing an additional transformer for cyclical features there that are omitted from this chapter.

We apply the transformers as follows to the `date` column of the DataFrame:

```
from sklearn.compose import ColumnTransformer
from sklearn.pipeline import Pipeline, make_pipeline
preprocessor = ColumnTransformer(
    transformers=[(
        "date",
        make_pipeline(
            DateFeatures(),
            ColumnTransformer(transformers=[
                ("cyclical", CyclicalFeatures(),
                 ["date_day", "date_weekday", "date_month"]
                )
            ], remainder="passthrough")
        ), ["date"],
    ),], remainder="passthrough"
)
```

The `remainder="passthrough"` argument is set in case we want to provide additional exogenous features for prediction.

We can define a pipeline of these preprocessing steps together with a model so that it can be fitted and applied to prediction:

```
from xgboost import XGBRegressor
pipeline = Pipeline(
```

```
    [
        ("preprocessing", preprocessor),
        ("xgb", XGBRegressor(objective="reg:squarederror", n_
estimators=1000))
    ]
)
```

The predictor is an XGBoost regressor. I didn't make much of an effort in terms of tweaking. The only parameter that we'll change is the number of estimators. We'll use an ensemble size (number of trees) of 1,000.

Now it's time to split the dataset into training and test sets. This includes two issues:

- We need to align the features with values ahead of time
- We need to split the dataset into two by a cutoff time

Let's first set the basic parameters for this. First, we want to predict into the future given a time horizon. Second, we need to decide how many data points we use for training and for testing:

```
TRAIN_SIZE = int(len(df) * 0.9)
HORIZON = 1
TARGET_COL = "new_cases"
```

We take 90% of points for training, and we predict 90 days into the future:

```
X_train, X_test = df.iloc[HORIZON:TRAIN_SIZE], df.iloc[TRAIN_
SIZE+HORIZON:]
y_train = df.shift(periods=HORIZON).iloc[HORIZON:TRAIN_SIZE][TARGET_
COL]
y_test = df.shift(periods=HORIZON).iloc[TRAIN_SIZE+HORIZON:][TARGET_
COL]
```

This does both the alignment and horizon. Therefore, we have the datasets for testing and training, both with features and labels that we want to predict with XGBoost.

Now we can train our XGBoost regression model to predict values within our HORIZON into the future based on the features we produced with our transformer and the current values.

We can fit our pipeline as follows:

```
FEATURE_COLS = ["date"]
pipeline.fit(X_train[FEATURE_COLS], y_train)
```

We can see the following pipeline parameters:

```
Pipeline(steps=[('preprocessing',
                ColumnTransformer(remainder='passthrough',
                                  transformers=[('date',
                                                Pipeline(steps=[('datefeatures',
                                                                DateFeatures()),
                                                               ('columntransformer',
                                                                ColumnTransformer(remainder='passthrough',
                                                                                  transformers=[('cyclical',
                                                                                                CyclicalFeatures(),
                                                                                                ['date_day',
                                                                                                 'date_weekday',
                                                                                                 'date_month'])])))]),
                                               ['date'])])),
               ('xgb',
                XGBRegressor(n_estimators=1000,
                             objective='reg:squarederror'))])
```

Figure 7.8: Pipeline parameters

If we create a series of the dates from beginning to end, we can get the predictions of the model for the whole time period:

```
MAX_HORIZON = 90
X_test_horizon = pd.Series(pd.date_range(
    start=df.date.min(),
    periods=len(df) + MAX_HORIZON,
    name="date"
)).reset_index()
```

The predict() method of the pipeline applied to X_test gives us the forecast:

```
forecasted = pd.concat(
    [pd.Series(pipeline.predict(X_test_horizon[FEATURE_COLS])),
pd.Series(X_test_horizon.date)],
    axis=1
)
forecasted.columns = [TARGET_COL, "date"]
```

We can do the same for the actual cases:

```
actual = pd.concat(
    [pd.Series(df[TARGET_COL]), pd.Series(df.date)],
    axis=1
)
actual.columns = [TARGET_COL, "date"]
```

Now, we can contrast the forecast with the actual values, y_test, in a plot:

```
fig, ax = plt.subplots(figsize=(12, 6))
forecasted.set_index("date").plot(linestyle='--', ax=ax)
```

```
actual.set_index("date").plot(linestyle='-.', ax=ax)
plt.legend(["forecast", "actual"])
```

This is the plot we are getting:

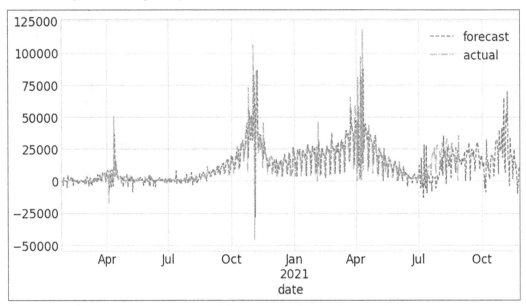

Figure 7.9: Forecast versus actual (XGBoost)

We can extract performance metrics over the test period like this:

```
from sklearn.metrics import mean_squared_error

test_data = actual.merge(forecasted, on="date", suffixes=("_actual",
"_predicted"))
mse = mean_squared_error(test_data.new_cases_actual, test_data.new_
cases_predicted, squared=False)  # RMSE
print("The root mean squared error (RMSE) on test set: {:.2f}".
format(mse))
```

We should be seeing something like this:

```
The root mean squared error (RMSE) on test set: 12753.41
```

Next, we'll create an ensemble model for time-series forecasting in Kats.

Ensembles with Kats

The Kats installation should be very easy in two steps. First, we install fbprophet, an old version of Facebook's Prophet library:

```
conda install -c conda-forge fbprophet
```

Now we install Kats with pip:

```
pip install kats
```

Alternatively, on Colab, we can install Kats like this:

```
!MINIMAL=1 pip install kats
!pip install "numpy==1.20"
```

We'll load the COVID cases dataset as before. Here's just the last line:

```
df = owid_covid[owid_covid.location == "France"].set_index("date",
drop=True).resample('D').interpolate(method='linear').reset_index()
```

We'll configure our ensemble model, fit it, and then do a forecast.

First, the configuration of our ensemble:

```
from kats.models.ensemble.ensemble import EnsembleParams,
BaseModelParams
from kats.models.ensemble.kats_ensemble import KatsEnsemble
from kats.models import linear_model, quadratic_model
model_params = EnsembleParams(
        [
                BaseModelParams("linear", linear_model.
LinearModelParams()),
                BaseModelParams("quadratic", quadratic_model.
QuadraticModelParams()),
        ]
    )
```

Here, we include only two different models, but we could have included other and more models, and we could have defined better parameters. This is an example only; for a more realistic exercise, which I leave to the reader, I'd suggest adding ARIMA and Theta models. We need to define hyperparameters for each forecasting model.

We also need to create ensemble parameters that define how the ensemble aggregate is to be calculated and how the decomposition should work:

```
KatsEnsembleParam = {
    "models": model_params,
    "aggregation": "weightedavg",
    "seasonality_length": 30,
    "decomposition_method": "additive",
}
```

To use a time-series with Kats, we have to convert our data from a DataFrame or series to a Kats time-series object. We can convert our COVID case data as follows:

```
from kats.consts import TimeSeriesData
TARGET_COL = "new_cases"

df_ts = TimeSeriesData(
    value=df[TARGET_COL], time=df["date"]
)
```

What is important for the conversion is the fact that Kats can infer the frequency of the index. This can be tested with `pd.infer_freq()`. In our case, `pd.infer_freq(df["date"])` should return D for daily frequency.

Now we can create our `KatsEnsemble` and fit it:

```
m = KatsEnsemble(
    data=df_ts,
    params=KatsEnsembleParam
).fit()
```

We can get separate predictions for each model using the `predict()` method. If we want to get the ensemble output, we have to call `aggregate()` after `predict()`:

```
m.predict(steps=90)
m.aggregate()
m.plot()
plt.ylabel(TARGET_COL)
```

We predict 90 days ahead. These predictions are stored as part of the model, so we don't need to capture the returned forecast. We can then aggregate the forecast from each model. Again, we don't need to get the returned DataFrame because this is stored inside the model object (`m.fcst_df`).

In the end, we plot the aggregated DataFrame using a Kats convenience function:

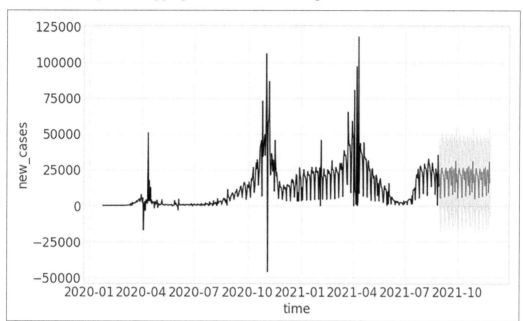

Figure 7.10: Kats ensemble model forecast

Since we can tweak this ensemble model by changing the base model parameter and adding new models, this can give us lots of room for improvement.

It's time to conclude this chapter with a summary of what we've learned.

Summary

In this chapter, we've discussed popular time-series machine learning libraries in Python. We then discussed and tried out a k-nearest neighbor algorithm with dynamic time warping for the classification of robotic failures. We talked about validation in time-series forecasting and we tried three different methods for forecasting COVID cases: Silverkite, Gradient Boosting with XGBoost, and ensemble models in Kats.

8

Online Learning
for Time-Series

In this chapter, we are going to dive into online learning and streaming data for time-series. Online learning means that we continually update our model as new data is coming in. The advantage of online learning algorithms is that they can handle the high speed and possibly large size of streaming data and are able to adapt to new distributions of the data.

We will discuss drift, which is important because the performance of a machine learning model can be strongly affected by changes to the dataset to the point that a model will become obsolete (stale).

We are going to discuss what online learning is, how data can change (drift), and how adaptive learning algorithms combine drift detection methods to adjust to this change in order to avoid the degradation of performance or costly retraining.

We're going to cover the following topics:

- Online learning for time-series
  - Online algorithms
- Drift
  - Drift detection methods
- Adaptive learning methods
- Python practice

We'll start with a discussion of online learning.

Online learning for time-series

There are two main scenarios of learning – online learning and offline learning. **Online learning** means that you are fitting your model incrementally as the data flows in (streaming data). On the other hand, **offline learning**, the more commonly known approach, implies that you have a static dataset that you know from the start, and the parameters of your machine learning algorithm are adjusted to the whole dataset at once (often loading the whole dataset into memory or in batches).

There are three major use cases for online learning:

- Big data
- Time constraints (for example, real time)
- Dynamic environments

Typically, in online learning settings, you have more data, and it is appropriate for big data. Online learning can be applied to large datasets, where it would be computationally infeasible to train over the entire dataset.

Another use case for online learning is where the inference and fitting are performed under time constraints (for example, a real-time application), and many online algorithms are very resource-efficient in comparison to offline algorithms.

A common application of online learning is on time-series data, and a particular challenge is that the underlying generating process of the time-series observations can change over time. This is called concept drift. While in the offline setting the parameters are fixed, in online learning, the parameters are continuously adapted based on new data. Therefore, online learning algorithms can deal with changes in the data, and some can deal with concept drift.

The table below summarizes some more differences between online and offline learning:

	Offline	Online
Necessity to monitor	Yes, models can become stale (the model will lose performance)	Adapt to changing data
Retraining costs	Expensive (from scratch)	Cheap (incremental)

Memory requirements	Possibly high memory demands	Low
Applications	Image classifiers, speech recognition, etc, where data is assumed to be static	Finance, e-commerce, economics, and health-care, where data is dynamically changing
Tools	tslearn, sktime, prophet	Scikit-Multiflow, River

Figure 8.1: Online vs offline learning methods in time-series

There are lots of other tools that are not specific to online learning but support online learning, such as the most popular deep learning libraries – PyTorch and TensorFlow, where models inherently support online learning and data loaders support streaming scenarios – through iterators, where data can be loaded in as needed.

A streaming formulation of a supervised machine learning problem can be posed as follows:

1. A data point $x \in \mathbb{R}^d$ is received at time t
2. The online algorithm predicts the label
3. The true label is revealed before the next data point comes in

In a batch setting, a set of n points $X \subset \mathbb{R}^d$ arrive all at once at time t, and all n points will be predicted by the online model before the true labels are revealed and the next batch of points arrives.

We can demonstrate the difference in Python code snippets to show the characteristic patterns of machine learning in online and offline settings. You should be familiar with offline learning, which looks like this for features X, target vector y, and model parameters params:

```
from sklearn import linear_model
offline_model = linear_model.LogisticRegression(params)
offline_model.fit(X, Y)
```

This should be familiar from previous chapters such as *Chapter 7, Machine Learning Models for Time-Series*. For simplicity, we are omitting data loading, preprocessing, cross-validation, and parameter tuning, among other issues.

Online learning follows this pattern:

```
from river import linear_model
online_model = linear_model.LogisticRegression(params)
for xi, yi in zip(X, y):
    online_model.learn_one(xi, yi)
```

Here, we are feeding point by point to the model. Again, this is simplified – I've omitted setting the parameters, loading the dataset, and more.

These snippets should make the main difference clear: learning on the whole dataset at once (offline) against learning on single points one by one (online).

I should mention evaluation methods for online methods:

- Holdout
- Prequential

In **Holdout**, we can apply the current model to the independent test set. This is popular in batch as well as online (stream) learning and gives an unbiased performance estimation.

In **Prequential Evaluation**, we test as we are going through the sequence. Each new data point is first tested and then trained on.

An interesting aspect of online learning is model selection, that is, how to select the best model among a set of candidate models. We looked at model selection for time-series models in *Chapter 4, Machine Learning Models for Time-Series*. There are different options for model selection in the online setting.

In a **Multi-Armed Bandit** (also **K-Armed Bandit**) problem, limited resources must be allocated between competing choices in a way that maximizes expected gain. Each choice ("arm") comes with its reward, which can be learned over time. Over time, we can adapt our preference for each of these arms and choose optimally in terms of expected reward. Similarly, by learning expected rewards for competing classification or regression models, methods for multi-armed bandits can be applied for model selection. In the practice section, we'll discuss multi-armed bandits for model selection.

In the following sections, we'll look at incremental methods and drift in more detail.

Online algorithms

Where data becomes available gradually over time or its size exceeds system memory limits, then incremental machine learning algorithms, whether supervised learning or unsupervised, can update parameters on parts of the data rather than starting the learning from scratch. **Incremental learning** is where parameters are continuously adapted to adjust a model to new input data.

Some machine learning methods inherently support incremental learning. Neural networks (as in deep learning), nearest neighbor, and evolutionary methods (for example, genetic algorithms) are incremental and can therefore be applied in online learning settings, where they are continuously updated.

Incremental algorithms may have random access to previous samples or prototypes (selected samples). These algorithms, such as based on the nearest neighbor algorithm, are called incremental algorithms with partial memory. Their variants can be suitable for cyclic drift scenarios.

Many well-known machine learning algorithms have incremental variants such as the adaptive random forest, the adaptive XGBoost classifier, or the incremental support vector machine.

Both reinforcement learning and active learning can be seen as types of online learning because they work in an online or active manner. We are going to discuss reinforcement learning in *Chapter 11, Reinforcement Learning for Time-Series*.

In online learning, updates are calculated continuously. At the heart of this is running statistics, so it could be illustrative to show how mean and variance can be calculated incrementally (in an online setting).

Let's look at the formulas for online arithmetic mean and online variance. As for the **online mean**, updating the mean μ_t at time point t can be done as follows:

$$\mu_{t+1} = \mu_t + \frac{x - \mu_t}{n_{t+1}},$$

where n_{t+1} is the number of previous updates – sometimes this is written as $t + 2$.

The **online variance** σ_t can be calculated based on the online mean and the running sum of squares s_t:

$$s_{t+1} = s_t + (x - \mu_t) \cdot (x - \mu_{t+1})$$

$$\sigma_{t+1} = \frac{s_{t+1}}{n_{t+1}}$$

A downside to offline algorithms is that they are sometimes more difficult to implement and that there's a learning curve to getting up to speed with libraries, algorithms, and methods.

scikit-learn, the standard library for machine learning in Python, only has a limited number of incremental algorithms. It is focused on batch-learning models. In contrast, there are specialized libraries for online learning with adaptive and incremental algorithms that cover many use cases, such as imbalanced datasets.

Research engineers, students, and machine learning researchers from the University of Waikato (New Zealand), Télécom ParisTech, and the École Polytechnique in Paris have been working on the **River library**. River is the result of merging two libraries: Creme (intended as a pun on incremental) and Scikit-Multiflow. River comes with many meta and ensemble methods. As a cherry on top, many of these meta or ensemble methods can use scikit-learn models as base models.

At the time of writing, the River library has 1,700 stars and implements many unsupervised and supervised algorithms. At the time of writing, the documentation of River is still a work in progress, but lots of functionality is available, as we'll see in the practical section at the end of this chapter.

This chart shows the popularity of River and Scikit-Multiflow over time (by the number of stars on GitHub):

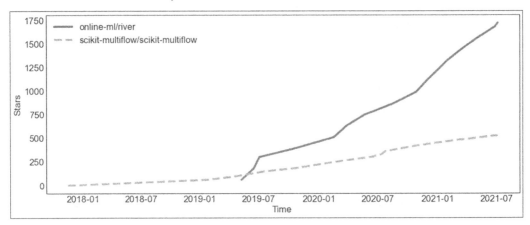

Figure 8.2: Star histories of the River and Scikit-Multiflow libraries

We can see that, while Scikit-Multiflow has risen steadily, this rise has been mostly flat. River overtook Scikit-Multiflow in 2019 and has continued to receive many star ratings from GitHub users. These star ratings are similar to a "like" on a social media platform.

This table shows a few online algorithms, some of which are suitable for drift scenarios:

Algorithm	Description
Very Fast Decision Tree (VFDT)	Decision tree made up of splits based on a few examples. Also called Hoeffding Tree. Struggles with drift.
Extremely Fast Decision Tree (EFDT)	Incrementally builds a tree by creating a split when confident and replaces the split if a better split is available. Assumes a stationary distribution.
Learn++.NSE	Ensemble of classifiers for incremental learning from non-stationary environments.

Figure 8.3: Online machine learning algorithms – some of them suitable for drift

The online algorithm par excellence is the **Hoeffding Tree** (Geoff Hulten, Laurie Spencer, and Pedro Domingos, 2001), also called **Very Fast Decision Tree** (VFDT). It is one of the most widely used online decision tree induction algorithms.

While some online learning algorithms are reasonably efficient, the attained performance can be highly sensitive to the ordering of data points, and potentially, they might never escape from a local minimum they ended up in, driven by early examples. Appealingly, VFDTs provide high classification accuracy with theoretical guarantees that they will converge toward the performance of decision trees over time. In fact, the probability that a VFDT and a conventionally trained tree will differ in their tree splits decreases exponentially with the number of examples.

The **Hoeffding bound**, proposed by Wassily Hoeffding in 1963, states that with probability $1 - \delta$, the calculated mean $\hat{z}$ of a random variable Z, calculated over n samples, deviates less than ϵ from the true mean $\bar{z}$:

$$|\bar{z} - \hat{z}| < \epsilon = \sqrt{\frac{R^2 \ln\left(\frac{1}{\delta}\right)}{2n}}$$

In this equation, R is the range of the random variable Z. This bound is independent of the probability distribution that is generating the observations.

As data comes in, new branches are continuously added and obsolete branches are cut out from the Hoeffding Tree. Problematically, however, under concept drift, some nodes may no longer satisfy the Hoefdding boundary.

In the next section, we'll look at drift, why you should care, and what to do about drift.

Drift

A major determinant of data quality is drift. **Drift** (also: **dataset shift**) means that the patterns in data change over time. Drift is important because the performance of a machine learning model can be adversely affected by changes to the dataset.

Drift transitions can occur abruptly, incrementally, gradually, or be recurring. This is illustrated here:

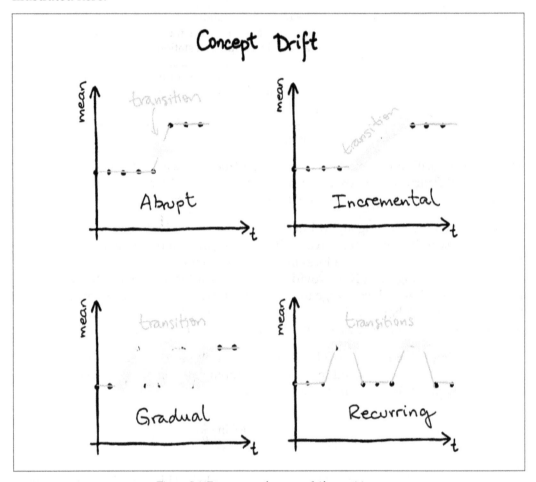

Figure 8.4: Four types of concept drift transitions

When the transition is abrupt, it happens from one time step to another without apparent preparation or warning. In contrast, it can also be incremental in the sense that there's first a little shift, then a bigger shift, then a bigger shift again.

When a transition happens gradually, it can look like a back and forth between different forces until a new baseline is established. Yet another type of transition is recurring or cyclical when there's a regular or recurring shift between different baselines.

There are different kinds of drift:

- Covariate drift
- Prior-probability drift
- Concept drift

Covariate drift describes a change in the independent variables (features). An example could be a regulatory intervention, where new laws would shake up the market landscape, and consumer behavior would follow different behaviors from before. An example would be if we want to predict chronic disease within 10 years given smoking behaviors, and smoking suddenly becomes much less prevalent, because of new laws. This means that our prediction could be less reliable.

Probability drift is a change in the target variable. An example could be that in fraud detection, the fraud incidence ratio changes; in retail, the average value of merchandise increases. One reason for drift could be seasonality – for example, selling more coats in winter.

In **concept drift**, a change occurs in the relationship between the independent and the target variables. The concept the term refers to is the relationship between independent and dependent variables. For example, if we wanted to predict the number of cigarettes smoked, we could assume that our model would become useless after the introduction of new laws. Please note that often, the term concept drift is applied in a broader sense as anything non-stationary.

Covariate drift: a change in the features $P(x)$.

Label drift (or **prior probability drift**): a change in the target variable $P(y)$.

Concept drift: (in supervised machine learning) changes in the conditional distribution of the target – in other words, the relationship between independent and dependent variables changes $P(y \mid X)$.

Commonly, when building machine learning models, we assume that points within different parts of the dataset belong to the same distribution.

While occasional anomalies, such as abnormal events, would usually be treated as noise and ignored, when there is a change in the distribution, models often have to be rebuilt from scratch based on new samples in order to capture the latest characteristics. This is the reason why we are testing time-series models with walk-forward validation as discussed in *Chapter 7, Machine Learning Models for Time-Series*. However, this training from scratch can be time-consuming and heavy on computing resources.

Drift causes problems for machine learning models since models can become stale – they become unreliable over time since the relationships they capture are no longer valid. This results in the performance degradation of these models. Therefore, approaches to forecasting, classification, regression, or anomaly detection should be able to detect and react to concept drift in a timely manner, so that the model can be updated as soon as possible. Machine learning models are often retrained periodically to avoid performance degradation happening. Alternatively, retraining can be triggered when needed based on either the performance monitoring of models or based on change detection methods.

As for applications on time-series, in many domains, such as finance, e-commerce, economics, and healthcare, the statistical properties of the time-series can change, rendering forecasting models useless. Puzzlingly, although the concept of the drift problem is well investigated in the literature, little effort has been invested in tackling it with time-series methods.

Gustavo Oliveira and others proposed in 2017 ("*Time-Series Forecasting in the Presence of Concept Drift: A PSO-based Approach*") training several time-series forecasting models. At each point in time, the parameters for each of these models were changed weighted by the latest performance (particle swarm optimization). When the best models (best particles) diverged beyond a certain confidence interval, retraining of models was triggered.

The charts below illustrate a combination of error-triggered retraining and online learning, one approach to time-series forecasting:

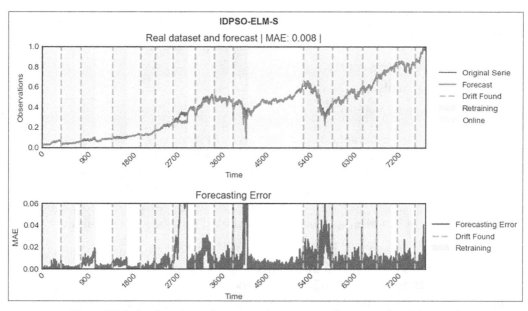

Figure 8.5: Online learning and retraining for time-series forecasting (IDPSO-ELM-S)

You can see the error rates increasing periodically as concept drift is occurring, and, where, based on the concept of drift detection, retraining is triggered.

Many online models have been specifically adapted to be robust to or handle concept drift. In this section, we'll discuss some of the most popular or best-performing ones. We'll also discuss methods for drift detection.

Drift detection methods

There are lots of different methods to explicitly detect drift and distributional changes in data streams. Page-Hinkley (Page, 1954) and Geometric Moving Average (Roberts, 2000) are two of the pioneers.

Drift detectors monitor the model performance usually through a performance metric, however, they can also be based on input features, although this is more of an exception. The basic idea is that when there is a change in the class distribution of the samples, the model does not correspond anymore to the current distribution, and the performance degrades (the error rate increases). Therefore, quality control of the model performance can serve as drift detection.

Drift detection methods can be categorized into at least three groups (after João Gama and others, 2014):

- Statistical process control
- Sequential analysis
- Windows-based comparison

Statistical process control methods take into account summary statistics such as the mean and standard deviation of model predictions. For example, the **Drift Detection Method (DDM**; João Gama and others, 2004) alerts if the error rate surpasses the previously recorded minimum error rate by three standard deviations. According to statistical learning theory, in a continuously trained model, errors should diminish with the number of samples, so this threshold should only be exceeded in the case of drift.

Sequential methods are based on thresholds of model predictions. For example, in the **Linear Four Rates** (Wang, 2015) method, the rates in the contingency table are updated incrementally. Significance is calculated according to a threshold that is estimated once at the start by Monte Carlo sampling. This method can handle class imbalance better than DDM.

 Contingency table: a table that compares the frequency distribution of variables. Specifically in machine learning classification, the table displays the predicted number of labels over the test set against the actual labels. In the case of binary classification, the cells show true positives, false positives, false negatives, and true negatives.

Windows-based approaches monitor the distribution of errors. For example, **ADWIN (ADaptive WINdowing)** was published by Albert Bifet and Ricard Gavaldà in 2007. Prediction errors within a time window W are partitioned into smaller windows, and the differences in mean error rates within these windows are compared to the Hoeffding bound. The original version proposes a variation of this strategy that has a time complexity of $O(log\ W)$, where W is the length of the window.

Some methods for drift detection are listed here:

Algorithm	Description	Type
Adaptive Windowing (ADWIN)	Adaptive sliding window algorithm based on thresholds.	Window-based
Drift Detection Method (DDM)	Based on the premise that the model's error rate should decrease over time.	Statistical
Early Drift Detection Method (EDDM)	Statistics over the average distance between two errors. Similar to DDM, but better for gradual drift.	Statistical
Hoffding's Drift Detection (HDDM)	Non-parametric method based on Hoeffding's bounds – either moving average-test or moving weighted average-test.	Window-based
Kolmogorov-Smirnov Windowing (KSWIN)	Kolmogorov-Smirnov test in windows of a time-series.	Window-based
Page-Hinkley	Statistical test for mean changes in Gaussian signals.	Sequential

Figure 8.6: Drift detection algorithms

Kolmogorov-Smirnov is a nonparametric test of the equality of continuous, one-dimensional probability distributions.

These methods can be used in the context of both regression and classification (and, by extension, forecasting). They can be used to trigger the retraining of models. For example, Hassan Mehmood and others (2021) retrained time-series forecasting models (among other models, they used Facebook's Prophet) if drift was detected.

Drift detectors all have their assumptions regarding input data. It is important to know these assumptions, and I've tried to outline these in the table, so you use the right detector with your dataset.

The drift detection methods listed above all incur a labeling cost. Since they all monitor the prediction results of a base classifier or an ensemble, they require that the class labels is available right after prediction. This constraint is unrealistic in some practical problems. There are other methods, not listed here, that can be based on anomaly detection (or novelty detection), feature distribution monitoring, or model-dependent monitoring. We saw a few of these methods in *Chapter 6, Unsupervised Methods for Time-Series*.

In the next section, we'll look at some methods that were designed to be resistant to drift.

Adaptive learning methods

Adaptive learning refers to incremental methods with drift adjustment. This concept refers to updating predictive models online to react to concept drifts. The goal is that by taking drift into account, models can ensure consistency with the current data distribution.

Ensemble methods can be coupled with drift detectors to trigger the retraining of base models. They can monitor the performance of base models (often with ADWIN) – underperforming models get replaced with retrained models if the new models are more accurate.

As a case in point, the **Adaptive XGBoost** algorithm (**AXGB**; Jacob Montiel and others, 2020) is an adaptation of XGBoost for evolving data streams, where new subtrees are created from mini-batches of data as new data becomes available. The maximum ensemble size is fixed, and once this size is reached, the ensemble is updated on new data.

In the Scikit-Multiflow and River libraries, there are several methods that couple machine learning methods with drift-detection methods, which regulate adaptation. Many of these were published by the maintainers of the two libraries. Here's a list with a few of these methods:

Algorithm	Description
K-Nearest Neighbors (KNN) classifier with ADWIN change detector	KNN with ADWIN change detector to decide which samples to keep or forget.

Adaptive Random Forest	Includes drift detectors per tree. It starts training in the background after a warning has been detected, and replaces the old tree if drift occurs.
Additive Expert ensemble classifier	Implements pruning strategies – the oldest or weakest base model will be removed.
Hoeffding Adaptive Tree (HAT)	Pairs ADWIN to detect drift and a Hoeffding Tree model to learn.
Very Fast Decision Rules	Similar to VFDT, but rule ensembles instead of a tree. In Scikit-Multiflow drift detection with ADWIN, DDM, and EDDM is supported.
Oza Bagging ADWIN	Instead of sampling with replacement, each sample is given a weight. In River, this can be combined with the ADWIN change detector.
Online CSB2	Online boosting algorithm that compromises between AdaBoost and AdaC2, and optionally uses a change detector.
Online Boosting	AdaBoost with ADWIN drift detection.

Figure 8.7: Adaptive learning algorithms

These methods are robust to drift by regulating the adaptation or learning with the concept of drift detection.

Let's try out a few of these methods!

Python practice

The installation in this chapter is very simple, since, in this chapter, we'll only use River. We can quickly install it from the terminal (or similarly from Anaconda Navigator):

```
pip install river
```

We'll execute the commands from the Python (or IPython) terminal, but equally, we could execute them from a Jupyter notebook (or a different environment).

Drift detection

Let's start off by trying out drift detection with an artificial time-series. This follows the example in the tests of the River library.

We'll first create an artificial time-series that we can test:

```
import numpy as np
np.random.seed(12345)
data_stream = np.concatenate(
    (np.random.randint(2, size=1000), np.random.randint(8, size=1000))
)
```

This time-series is composed of two series that have different characteristics. Let's see how quickly the drift detection algorithms pick up on this.

Running the drift detector over this means iterating over this dataset and feeding the values into the drift detector. We'll create a function for this:

```
def perform_test(drift_detector, data_stream):
    detected_indices = []
    for i, val in enumerate(data_stream):
        in_drift, in_warning = drift_detector.update(val)
        if in_drift:
            detected_indices.append(i)
    return detected_indices
```

Now we can try the ADWIN drift detection method on this time-series. Let's create another method to plot the drift points overlaid over the time-series:

```
import matplotlib.pyplot as plt

def show_drift(data_stream, indices):
    fig, ax = plt.subplots(figsize=(16, 6))
    ax.plot(data_stream)
    ax.plot(
        indices,
        data_stream[indices],
        "ro",
        alpha=0.6,
        marker=r'$\circ$',
        markersize=22,
        linewidth=4
    )
    plt.tight_layout()
```

This is the plot for the ADWIN drift points:

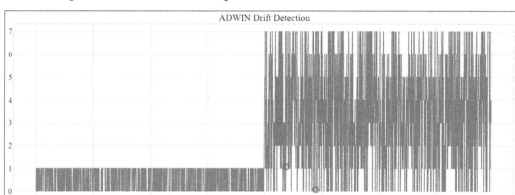

Figure 8.9: ADWIN drift points on our artificial dataset

I'd encourage you to play around with this and to also try out the other drift detection methods.

Next, we'll do a regression task.

Regression

We are going to estimate the occurrence of medium-class solar flares.

For this, we'll use the solar flares dataset from the UCI machine learning repository. The River library ships with a zipped column-separated dataset of the dataset, and we'll load this, specify the column types, and choose the outputs we are interested in.

Let's plot the ADWIN results now:

```
from river import stream
from river.datasets import base


class SolarFlare(base.FileDataset):
    def __init__(self):
        super().__init__(
            n_samples=1066,
            n_features=10,
            n_outputs=1,
            task=base.MO_REG,
            filename="solar-flare.csv.zip",
```

```
            )

    def __iter__(self):
        return stream.iter_csv(
            self.path,
            target="m-class-flares",
            converters={
                "zurich-class": str,
                "largest-spot-size": str,
                "spot-distribution": str,
                "activity": int,
                "evolution": int,
                "previous-24h-flare-activity": int,
                "hist-complex": int,
                "hist-complex-this-pass": int,
                "area": int,
                "largest-spot-area": int,
                "c-class-flares": int,
                "m-class-flares": int,
                "x-class-flares": int,
            },
        )
```

Please note how we are choosing the number of targets and the converters, which contain the types for all feature columns.

Let's have a look at what this looks like:

```
from pprint import pprint
from river import datasets

for x, y in SolarFlare():
    pprint(x)
    pprint(y)
    break
```

We see the first point of the dataset (the first row of the dataset):

```
                    {'activity': 1,
                     'area': 1,
                     'c-class-flares': 0,
                     'evolution': 3,
                     'hist-complex': 1,
                     'hist-complex-this-pass': 1,
                     'largest-spot-area': 1,
                     'largest-spot-size': 'A',
                     'previous-24h-flare-activity': 1,
                     'spot-distribution': 'X',
                     'x-class-flares': 0,
                     'zurich-class': 'H'}
                  0
```

Figure 8.10: First point of the solar flare dataset for medium-sized flares

We see the ten feature columns as a dictionary and the output as a float.

Let's build our model pipeline in River:

```
import numbers
from river import compose
from river import preprocessing
from river import tree

num = compose.SelectType(numbers.Number) | preprocessing.MinMaxScaler()
cat = compose.SelectType(str) | preprocessing.
OneHotEncoder(sparse=False)
model = tree.HoeffdingTreeRegressor()
pipeline = (num + cat) | model
```

A pipeline like this is very pleasant to read: numeric features get min-max scaled, while string features get one-hot encoded. The preprocessed features get fed into a Hoeffding Tree model for regression.

We can now learn our model prequentially, by predicting values and then training them as discussed before:

```
from river import evaluate
from river import metrics

metric = metrics.MAE()
evaluate.progressive_val_score(SolarFlare(), pipeline, metric)
```

We are using the **Mean Absolute Error (MAE)** as our metric.

We get an MAE of 0.096979.

This prequential evaluation `evaluate.progressive_val_score()` is equivalent to the following:

```
errors = []
for x, y in SolarFlare():
    y_pred = pipeline.predict_one(x)
    metric = metric.update(y, y_pred)
    errors.append(metric.get())
    pipeline = pipeline.learn_one(x, y)
```

I've added two extra lines to collect the error over time as the algorithm learns.

Let's plot this:

```
fig, ax = plt.subplots(figsize=(16, 6))
ax.plot(
    errors,
    "ro",
    alpha=0.6,
    markersize=2,
    linewidth=4
)
ax.set_xlabel("number of points")
ax.set_ylabel("MAE")
```

This plot shows how this error evolves as a function of the number of points the algorithm encounters:

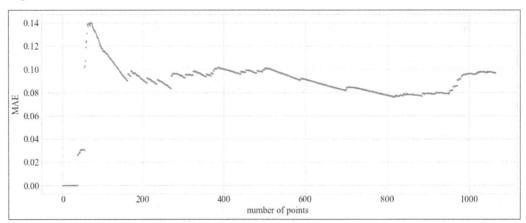

Figure 8.11: MAE by the number of points

We can see that, after 20-30 points, after the metric stabilizes, the Hoeffding Tree starts learning and the error keeps decreasing until about 800 points, at which point the error increases again. This could be a row ordering effect.

A dataset that has concept drift is the use case for an adaptive model. Let's compare adaptive and non-adaptive models on a dataset with concept drift:

```
from river import (
    synth, ensemble, tree,
    evaluate, metrics
)


models = [
    tree.HoeffdingTreeRegressor(),
    tree.HoeffdingAdaptiveTreeRegressor(),
    ensemble.AdaptiveRandomForestRegressor(seed=42)
]
```

We will compare the Hoeffding Tree Regressor, the Adaptive Hoeffding Tree Regressor, and the Adaptive Random Forest Regressor. We take the default settings for each model.

We can use a synthetic dataset for this test. We can train each of the aforementioned models on the data stream and look at the **Mean Squared Error (MSE)** metric:

```
for model in models:
    metric = metrics.MSE()
    dataset = synth.ConceptDriftStream(
        seed=42, position=500, width=40
    ).take(1000)
    evaluate.progressive_val_score(dataset, model, metric)
    print(f"{str(model.__class__).split('.')[-1][:-2]}: {metric.get():e}")
```

The `evaluate.progressive_val_score` method iterates over each point of the dataset and updates the metric. We get the following result:

```
HoeffdingTreeRegressor: 8.427388e+42
HoeffdingAdaptiveTreeRegressor: 8.203782e+42
AdaptiveRandomForestRegressor: 1.659533037987239+42
```

Your results might vary a bit because of the nature of these algorithms. We could set a random number generator seed to avoid this, however, I found it worth emphasizing this point.

We see the model error (MSE) in scientific notation, which helps in understanding the numbers, since they are quite large. You see the errors expressed in two parts, first a factor and then the order of magnitude as exponents to ten. The orders of magnitudes are the same for the three models, however, the Adaptive Random Forest Regressor obtained about a fifth of the error of what the other two got.

We can also visualize the error over time as the models learn and adapt:

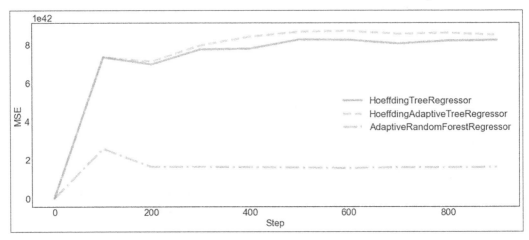

Figure 8.12: Model performance for a concept drift data stream (MSE)

There's no non-adaptive version of the random forest algorithm in River, so we can't compare this. We can't draw a clear conclusion about whether adaptive algorithms actually work better.

There are lots of other models, meta models, and preprocessors to try out if you want to have a play around.

Model selection

We've mentioned model selection with multi-armed bandits earlier in this chapter, and here we'll go through a practical example. This is based on documentation in River.

Let's use UCBRegressor to select the best learning rate for a linear regression model. The same pattern can be used more generally to select between any set of (online) regression models.

First, we define the models:

```
from river import compose
from river import linear_model
```

```
from river import preprocessing
from river import optim
models = [
    compose.Pipeline(
        preprocessing.StandardScaler(),
        linear_model.LinearRegression(optimizer=optim.SGD(lr=lr))
    )
    for lr in [1e-4, 1e-3, 1e-2, 1e-1]
]
```

We build and evaluate our models on the TrumpApproval dataset:

```
from river import datasets
dataset = datasets.TrumpApproval()
```

We'll apply the UCB bandit, which calculates reward for regression models:

```
from river.expert import UCBRegressor
bandit = UCBRegressor(models=models, seed=1)
```

The bandit provides methods to train its models in an online fashion:

```
for x, y in dataset:
    bandit = bandit.learn_one(x=x, y=y)
```

We can inspect the number of times (as a percentage) each arm has been pulled.

```
for model, pct in zip(bandit.models, bandit.percentage_pulled):
    lr = model["LinearRegression"].optimizer.learning_rate
    print(f"{lr:.1e} — {pct:.2%}")
```

The percentages for the four models are as follows:

```
1.0e-04 — 2.45%
1.0e-03 — 2.45%
1.0e-02 — 92.25%
1.0e-01 — 2.85%
```

We can also look at the average reward of each model:

```
for model, avg in zip(bandit.models, bandit.average_reward):
    lr = model["LinearRegression"].optimizer.learning_rate
    print(f"{lr:.1e} — {avg:.2f}")
```

The reward is as follows:

```
1.0e-04 — 0.00
1.0e-03 — 0.00
1.0e-02 — 0.74
1.0e-01 — 0.05
```

We can also plot the reward over time as it gets updated based on model performance:

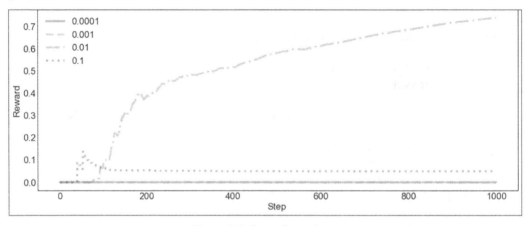

Figure 8.13: Reward over time

You can see that the rewards slowly become known as we step through the data and the models get updated. The model rewards clearly separate at around 100 time steps, and at around 1,000 time steps, seem to have converged.

We can also plot the percentage of the time each of the different models have been chosen at each step (this is based on the reward):

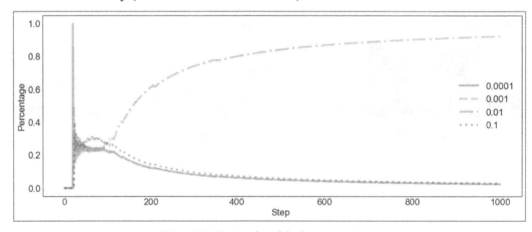

Figure 8.14: Ratios of models chosen over time

This distribution roughly follows the reward distribution over time. This should be expected since the model choice depends on reward (and a random number that regulates exploration).

We can also select the best model (the one with the highest average reward).

```
best_model = bandit.best_model
```

The learning rate chosen by the bandit is:

```
best_model["LinearRegression"].intercept_lr.learning_rate
```

The learning rate is 0.01.

Summary

In this chapter, we've discussed online learning. We've talked about some of the advantages of online learning methods:

- They are efficient and can handle high-speed throughput
- They can work on very large datasets
- And they can adjust to changes in data distributions

Concept drift is a change in the relationship between data and the target to learn. We've talked about the importance of drift, which is that the performance of a machine learning model can be strongly affected by changes to the dataset to the point that a model will become obsolete (stale).

Drift detectors don't monitor the data itself, but they are used to monitor model performance. Drift detectors can make stream learning methods robust against concept drift, and in River, many adaptive models use a drift detector for partial resets or for changing learning parameters. Adaptive models are algorithms that combine drift detection methods to avoid the degradation of performance or costly retraining. We've given an overview of a few adaptive learning algorithms.

In the Python practice, we've played around with a few of the algorithms in the River library, including drift detection, regression, and model selection with a multi-armed bandit approach.

9

Probabilistic Models for Time-Series

Probability is a measure of how likely something is to occur. In sales forecasting, an estimate of uncertainty is crucial because these forecasts, providing insights into cash flow, margin, and revenue, drive business decisions on which depend the financial stability and the livelihoods of employees. This is where probabilistic models for time-series come in. They help us make decisions when an estimate of certainty is important.

In this chapter, I'll introduce Prophet, Markov models, and Fuzzy time-series models. At the end, we'll go through an applied exercise with these methods.

Another application for probabilistic modeling is estimating counterfactuals, where we can estimate treatment effects in experiments. We'll discuss the concept of Bayesian Structural Time-Series Models, and we'll run through a practical example with a time-series in the practice section.

We're going to cover the following topics:

- Probabilistic Models for Time-Series
- Prophet
- Markov Models
- Fuzzy Modeling
- Bayesian Structural Time-Series Models

- Python Exercise:
  - Prophet
  - Markov Switching Model
  - Fuzzy Time-Series
  - Bayesian Structural Time-Series Models

We'll start with an introduction to probabilistic time-series predictions.

Probabilistic Models for Time-Series

As mentioned in the introduction, probabilistic models can help us make decisions under uncertainty, and in situations where estimates have to come with quantified confidence, such as in financial forecasting, this can be crucial. For predictions of sales or cash flow, attaching probabilities to model predictions can make it easier for financial controllers and managers to act on the new information.

Some well-known algorithms include Prophet, explicitly designed for monitoring operational metrics and **key performance indicators** (**KPIs**), and Markov models. Others are stochastic deep learning models such as **DeepAR** and **DeepState**. Since we are dealing with deep learning models in *Chapter 10, Deep Learning Models*, we'll not deal with them in detail in this chapter.

The Prophet model comes from Facebook (Taylor and Letham, 2017) and is based on a decomposable model with interpretable parameters. A guiding design principle was that parameters can be intuitively adjusted by analysts.

Both Prophet and the Silverkite algorithm, which we introduced in *Chapter 7, Machine Learning Models for Time-Series*, aim for accurate predictions with time-series that can have changing trends, seasonality, and recurring events (such as holidays), and short-term effects, and are therefore well suited for many applications in data science, where the focus is on tasks such as resource planning, optimizing financial decisions, and tracking progress for operational analysis – typical tasks for operations research.

Other types of models of particular interest within the application of time-series include Markov models, which we'll discuss in a dedicated section.

Bayesian Structural Time-Series (**BSTS**) models, which we mentioned in *Chapter 6, Unsupervised Models for Time-Series*, allow the quantification of the posterior uncertainty of the individual components, control the variance of the components, and impose prior beliefs on the model. The BSTS model is a technique that can be used for feature selection, time-series forecasting, and inferring causal relationships. This last point, causal inference, is another use case for probabilistic models in time-series. Understanding the impact of interventions can be important, for example, with A/B tests.

The following plot illustrates the popularity of a few selected probabilistic libraries suitable for time-series predictions:

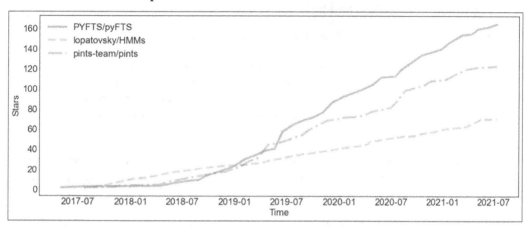

Figure 9.1: Libraries for the probabilistic modeling of time-series

You can see that, of these three libraries, pyFTS outranks the other two. I haven't included the statsmodels library, which includes a few probabilistic models. Neither have I included Prophet. Both statsmodels and Prophet would have outstripped HMMs by far, a library for Hidden Markov Models, and Pints, a library for noisy time-series.

Neither have I included neural network or deep learning libraries such as TensorFlow Probability or Gluon-TS. Deep learning will be the topic of *Chapter 10, Deep Learning for Time-Series*.

Let's start with a forecasting model in Prophet!

Prophet

Facebook's Prophet is both a Python/R library and the algorithm that comes with it. The algorithm was published in 2017 ("Forecasting at Scale" by Sean Taylor and Benjamin Letham). The authors write that the problems of forecasting and anomaly detection in practice involve the complexity of handling a variety of idiosyncratic forecasting problems at Facebook with piecewise trends, multiple seasonalities, and floating holidays, and building trust across the organization in these forecasts.

With these goals in mind, Prophet was designed to be scalable to many time-series, flexible enough for a wide range of business-relevant, possibly idiosyncratic time-series, and at the same time intuitive enough to be configurable by domain experts who might have little knowledge of time-series methods.

The Prophet algorithm is similar to the **Generalized Additive Model (GAM)** and formalizes the relationship between the forecast for the three model components, trend (growth), seasonality, and holiday, as follows:

$$y(t) = g(t) + s(t) + h(t) + \epsilon_t$$

The error term epsilon represents the residual — idiosyncratic changes not accommodated by the model. All functions use time as the regressor. The three effects are additive; however, Sean Taylor and Benjamin Letham advise that multiplicative seasonality, where the seasonal effect is a factor that multiplies g(t), can be accomplished through a log transform.

The trend or growth function can be linear or logistic for saturating growth. Both can incorporate piecewise effects through change points. The seasonality models periodic effects based on the Fourier series.

Change point selection in Prophet is automated. Parameters are optimized via the Broyden–Fletcher–Goldfarb–Shanno (BFGS) algorithm, as implemented in the Stan platform for statistical modeling.

Probabilistic models bring the advantage of providing a measure of certainty with the prediction; however, their predictions are not necessarily better than those of non-probabilistic models. Benchmark results of Prophet against other models have seen mixed results.

In their 2020 paper "*A Worrying Analysis of Probabilistic Time-Series Models for Sales Forecasting*", Seungjae Jung and others validated probabilistic time-series models on a large-scale dataset. The univariate time-series consists of the daily sales from an e-commerce website.

They compared two deep-learning probabilistic models, DeepAR and DeepState, and Prophet to baseline models that comprised a **moving average (MA)**, **linear regression (LR)**, a **multi-layer perceptron (MLP)**, and **Seasonal ARIMA (SARIMA)**. You should remember from *Chapter 5, Moving Averages and Autoregressive Models*, that the MA is a simple unweighted mean of preceding days. They tried 72 different hyperparameters for prophet and all baseline models.

They found that probabilistic models in their test failed to outperform even the simplest baseline models such as MLP and LR in terms of **root mean squared error (RMSE)** and **mean absolute percentage error (MAPE)**. Overall, Prophet performed the worst of all models. As always, model performance depends on the dataset and the task at hand — there's no silver bullet.

Let's see how Markov Models work!

Markov Models

A Markov chain is a probabilistic model describing a sequence of possible events that satisfies the Markov property.

Markov property: In a sequence or stochastic process that possesses the Markov property, the probability of each event depends only on the immediately preceding state (rather than earlier states). These sequences or processes can also be called **Markovian**, or a **Markov Process**.

Named after Russian mathematician Andrey Markov, the Markov property is very desirable since it significantly reduces the complexity of a problem. In forecasting, instead of taking into account all previous states, t-1, t-2, ..., 0, only t-1 is considered.

Similarly, the **Markov assumption**, for a mathematical or machine learning model is that the sequence satisfies the Markov property. In models such as the Markov chain and Hidden Markov model, the process or sequence is assumed to be a Markov process.

In a **discrete-time Markov chain** (**DTMC**), the sequence (or chain) transitions between states at discrete time steps. Markov chains can also be operating at continuous time steps. This less common model is called a **continuous-time Markov chain** (**CTMC**).

In a **hidden Markov model** (**HMM**), it is assumed that the process X follows unobservable states Y, which is another process whose behavior depends on X. The HMM models this latent or hidden process Y based on X.

Yet another Markov-type model is the nonlinear regime-switching model (also: Markov switching model). Invented by James Hamilton in 1989, the regime-switching model specifically addresses situations of abrupt changes, where more conventional linear models would struggle to capture distinct behaviors. The regime-switching model is an autoregressive model, where the mean of the process switches between regimes.

For the practical example, we'll follow a `statsmodels` library implementation and build a model to replicate Hamilton's 1989 model. Hamilton modeled a time-series of the real gross national product (RGNP), a macroeconomic measure of the value of economic output adjusted for price changes, between 1951 and 1984.

We'll use a Markov switching model of order 4, which can be written as follows:

$$y_t = \mu_{S_t} + \phi_1\left(y_{t-1} - \mu_{S_{t-1}}\right) + \phi_2\left(y_{t-2} - \mu_{S_{t-2}}\right) + \phi_3\left(y_{t-3} - \mu_{S_{t-3}}\right) + \phi_4\left(y_{t-4} - \mu_{S_{t-4}}\right) + \epsilon_t$$

For each state (or period), the regime transitions according to the following matrix of transition probabilities:

$$P(S_t = s_t | S_{t-1}) = \begin{pmatrix} p_{00}\ p_{10} \\ p_{01}p_{11} \end{pmatrix}$$

p_{ij} is the probability of transitioning from regime i to regime j.

In this case, we are modeling two regimes.

In the next section, we'll discuss a Fuzzy approach to time-series modeling.

Fuzzy Modeling

Fuzzy logic and fuzzy set theory were developed by Lotfi Zadeh in the 1960s and 70s while a professor at the University of California, Berkeley. Born to Persian and Jewish Russian parents in Baku, Azerbaijan, he completed his schooling in Tehran, Iran, and later moved to the USA, where he studied at MIT and Columbia. As a result, he was familiar with how concepts are understood in different cultures and expressed in different languages. This inspired his research approach to approximate reasoning and linguistic variables that he formalized as fuzzy theory.

Fuzzy set theory is an approach that can deal with problems relating to ambiguous, subjective, and imprecise judgments. Vagueness is inherent in everyday language, and fuzziness was invented to express this and work with it in an intuitive manner. Fuzzy logic expresses subjective belief and vagueness. It can and has been claimed that probability theory is a subset of fuzzy logic.

Fuzzy sets are sets whose elements have degrees of membership. In Fuzzy logic, instead of binary (Boolean) truth values, True and False, the unit interval [0, 1] is used as a basis for rules of inference. More formally, the membership function, which expresses the degree of certainty that an element belongs to the set, is characterized by a membership mapping function:

$$\mu_A \colon X \to [0,1],$$

For example, a well-known algorithm, **fuzzy c-means** (James Bezdek, 1981), based on k-means, returns degrees of membership to clusters. This means that each point can belong to each cluster, but to varying degrees. This fuzzy set membership is in contrast to the so-called crisp partitioning typically returned by other clustering algorithms, where a point is either a member of a cluster or not.

For fuzzy logic, all set operations, such as equality, sub- and superset, union, and intersection, had to be redefined. The union between two fuzzy sets (or relations) is defined as the max operation on each point, while the intersection is defined as the min operation. More formally, the union between two sets A and B, $A \cup B$, is defined as:

$$\forall x \in U: \mu_{A \cup B}(x) = max\big(\mu_A(x), A(x), \mu_B(x)\big),$$

where $\mu_A(x)$ is the membership function for point x.

Song and Chissom (1993) proposed a first-order, time-invariant fuzzy time-series model to forecast enrollments at the University of Alabama. This is formalized as follows:

$$A_t = A_{t-1} \circ R,$$

where A_t is the enrollment in year t, R is the union of fuzzy relations, and ∘ is the fuzzy Max-Min composition operator.

The **Max-Min composition operation**, $C = A \circ B$, is obtained by taking the minimum term-by-term of the ith row of A and the jth column of B, and taking the maximum of these n minimums.

This is illustrated in the diagram below (from the Matrix Multiplication page on Wikipedia):

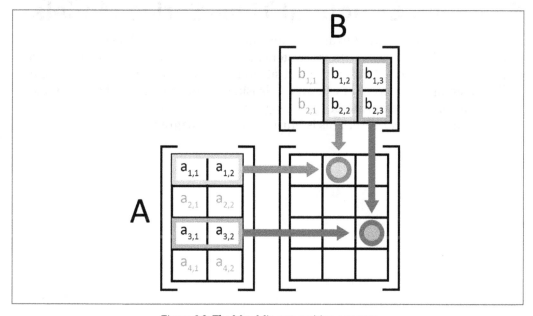

Figure 9.2: The Max-Min composition operator

The values at the positions marked with circles are calculated as follows:

$$c_{1,2} = \max\left(\min\left(a_{1,1}, b_{1,2}\right), \left(\min\left(a_{1,2}, b_{2,2}\right)\right.\right.$$

$$c_{3,3} = \max\left(\min\left(a_{3,1}, b_{1,3}\right), \left(\min\left(a_{3,2}, b_{2,3}\right)\right.\right.$$

In Song and Chissom's approach, relations between values at time t and values preceding it are extracted and carried forward for the forecast. A necessary preprocessing step in their algorithm is the conversion of time-series X into a fuzzy time-series Y. This is called fuzzification and consists of constraining an input from the set of real values to fuzzy memberships of a discrete set. This quantization can be performed by vector quantization methods such as Kohonen Self-Organizing Maps (SOM), an unsupervised machine learning method that produces a low-dimensional representation.

While fuzzy time-series models have not enjoyed widespread application, they have been shown to be competitive in some applications to more traditional approaches, such as SARIMA (for example, Maria Elena et al., 2012). They work on discrete and continuous time-series and produce interpretable models for forecasting.

In the following section, we'll do a few practice examples for probabilistic time-series predictions.

Bayesian Structural Time-Series Models

In causal inference, we want to analyze the effect of a treatment. The treatment can be any action that interacts with the system or environment that we care about, from changing the colors of a button on a website to the release of a product. We have the choice of taking the action (for example, releasing the product), thereby observing the outcome under treatment, or not taking the action, where we observe the outcome under no treatment. This is illustrated in the diagram here:

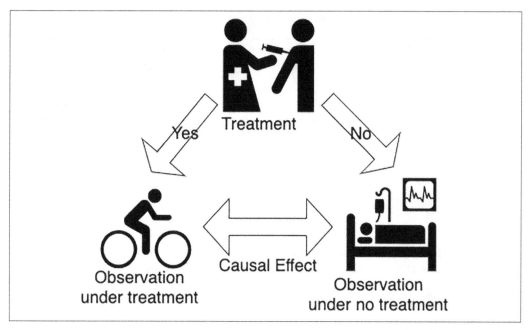

Figure 9.3: Causal effect of a treatment

In the diagram, an action is taken or not (medicine is administered to a patient), and depending on whether the action is taken we see the patient recovering (cycling) or going into intensive care.

A causal effect is the difference between what happens under treatment and what happens under no treatment. The problem with this is that we can't observe both potential outcomes at the same time.

However, we can run an experiment to observe both potential outcomes under treatment and potential outcomes under no treatment, such as in an A/B test, where the treatment is given only to a subset of the total population and the treatment condition B can be compared against the control condition A.

We can tabulate potential outcomes like this:

Unit	Treatment status, T_i	Outcome under treatment, $Y_i(1)$	Outcome under no treatment, $Y_i(0)$	Covariates, X_i
1	1	✓	estimate	✓
2	1	✓	estimate	✓
3	0	estimate	✓	✓
4	0	estimate	✓	✓

Figure 9.4: Potential outcomes from an experiment

In the first column, *Unit*, we see the sample indexes. Each row refers to a separate unit or sample in the population. The second column (*Treatment status*) encodes if the treatment was administered (1) or not (0). In the third and fourth columns, *Outcome under treatment* and *Outcome under no treatment*, respectively, are registered.

The marks show what should be obvious: when there is a treatment, we can observe the outcomes under treatment, but not the outcomes under no treatment. Inversely, when there is no treatment, we can observe the outcomes under no treatment, but not the outcomes under treatment.

Finally, in the last column, there are additional variables that can help us in our model that are available irrespective of treatment or no treatment.

With **Bayesian Structural Time-Series (BSTS)**, the focus is on estimating the treatment effect in the absence of an experiment. We can estimate or impute the counterfactuals, which are the unknown potential outcomes of an experiment. This allows to compare the outcomes under treatment against outcomes under no treatment, and therefore to quantify the causal treatment effect.

The model consists of three main components:

1. Kalman filter
2. Variable selection
3. Bayesian model averaging

Kalman filters are used for time-series decomposition. This allows the modeling of trends, seasonality, and holidays. In the Bayesian variable selection step (the spike-and-slab technique), the most important regression predictors are selected. Finally, in the model averaging, the prediction results are combined.

The application of BSTS models for change point and anomaly detection was described in the paper *"Predicting the Present with Bayesian Structural Time-Series"* (2013), by Steven L. Scott and Hal Varian.

A paper outlining the application of BSTS for estimating the causal effect of interventions was published in 2015 by Google Research (*"Inferring causal impact using Bayesian structural time-series models"*, by Kay H. Brodersen, Fabian Gallusser, Jim Koehler, Nicolas Remy, and Steven L. Scott).

The mathematical details of this are beyond the scope of this chapter. Fortunately, we can use a Python library to apply BSTS models. We'll run through an example in a practice section of this chapter.

We can practice now some of the theory that we have learned so far in this chapter.

Python Exercise

Let's put into practice what we've learned in this chapter so far. We'll be doing a model in Prophet, a Markov Switching model, a Fuzzy time-series model, and a BSTS model.

Let's get started with Prophet!

Prophet

First, let's make sure we have everything installed that we need. Let's quickly install the required libraries. We can do this from the terminal (or similarly from the Anaconda navigator):

```
pip install -U pandas-datareader plotly
```

You'll need a recent version of pandas-datareader, otherwise you might get a `RemoteDataError`.

We'll use the Prophet model through Facebook's Prophet library. Let's install it:

```
pip install prophet
```

Once this is done, we are set to go.

In this example, we'll use the daily Yahoo closing stock values in this chapter that we used in several examples in *Chapter 7, Machine Learning Models for Time-Series*.

To recap, we can download the daily Yahoo stock history from 2001 to 2021 in pandas-datareader as follows:

```
import pandas as pd
import numpy as np
from pandas_datareader.data import DataReader
from datetime import datetime
yahoo_data = DataReader('JPM', 'yahoo', datetime(2001,6,1),
datetime(2021,6,1))
yahoo_df = yahoo_data['Adj Close'].to_frame().reset_index('Date')
```

This gives us a pandas DataFrame with two columns, the adjusted daily closing value and the date. Let's quickly check the datatypes of these two columns:

```
yahoo_df.dtypes
```

These are the datatypes:

```
Date          datetime64[ns]
Adj Close           float64
dtype: object
```

The `Date` column is datetime in nanoseconds. `Adj Close` is of type float.

We'll feed this into the `fit()` method for training:

```
from prophet import Prophet

forecaster = Prophet()
forecaster.fit(
    yahoo_df.rename(columns={"Date": "ds", "Adj Close": "y"})
)
```

We have to rename our columns ds and y in order to stick to the Prophet conventions. We have a trained Prophet model now.

We'll then create a new DataFrame that will have future dates. We'll be able to stick this DataFrame into the `predict()` method of the Prophet model:

```
future = forecaster.make_future_dataframe(periods=90)
```

The forecast is calling the `predict()` method with this new DataFrame:

```
forecast = forecaster.predict(future)
```

The `forecast` DataFrame contains the upper and lower confidence intervals alongside the forecast. The `ds` columns is the date corresponding to the forecast.

Let's plot the forecasts against the actual data:

```
forecaster.plot(forecast, figsize=(12, 6));
```

Here's the plot:

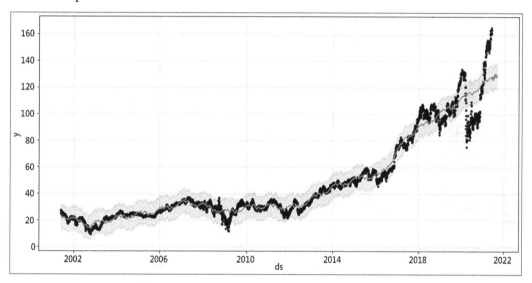

Figure 9.5: Forecast versus actual time-series (Prophet)

You might want to compare this plot to the one in *Chapter 7, Machine Learning Models for Time-Series*. The actual data is thick and bold, while the forecast is thinner. The upper and lower confidence intervals are around the forecast.

We can inspect the forecasts by looking at the DataFrame:

	ds	yhat	yhat_lower	yhat_upper
5116	2021-08-26	127.904392	119.437334	136.699758
5117	2021-08-27	127.923486	119.152098	137.186791
5118	2021-08-28	129.850793	121.337176	139.016644
5119	2021-08-29	129.898472	121.190163	138.472808
5120	2021-08-30	128.023878	119.387225	137.481971

Figure 9.6: Table of forecasts (Prophet)

It is quite easy to get a first model, and there are many ways to tweak it.

Markov Switching Model

For the Markov Switching model, we'll use the `statsmodels` library. If you don't have it installed yet, you can install it like this:

```
pip install statsmodels
```

We'll use a dataset with `statsmodels` in this example. This is based on the `statsmodels` tutorial on Markov switching autoregression models. We can get the dataset from the Stata Press publishing house on their website:

```
from statsmodels.tsa.regime_switching.tests.test_markov_autoregression
import statsmodels.api as sm
import seaborn as sn
import pandas as pd
dta = pd.read_stata('https://www.stata-press.com/data/r14/rgnp.dta').
iloc[1:]
dta.index = pd.DatetimeIndex(dta.date, freq='QS')
dta_hamilton = dta.rgnp
```

This gives us a pandas series of the RGNP, and the index annotates the dates. Let's quickly plot this:

```
dta_hamilton.plot(title='Growth rate of RGNP')
```

We get the following plot:

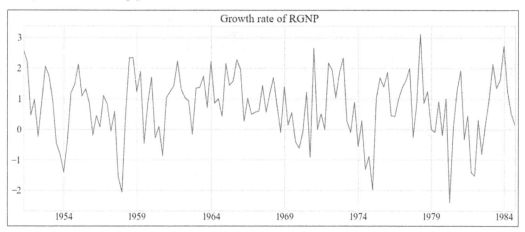

Figure 9.7: Growth rate of RGNP

We'll model domestic recessions and expansions. The model will include transition probabilities between these two regimes and predict probabilities of expansion or recession at each time point.

Let's fit the 4th order Markov switching model. We'll specify two regimes:

```
import statsmodels.api as sm
mod_hamilton = sm.tsa.MarkovAutoregression(dta_hamilton, k_regimes=2,
order=4, switching_ar=False)
res_hamilton = mod_hamilton.fit()
```

We now have the model fitted via maximum likelihood estimation to the RGNP data. We've set `switching_ar=False` because the `statsmodels` implementation defaults to switching autoregressive coefficients.

Let's have a look at the `statsmodels` model summary:

```
print(res_hamilton.summary())
```

We get the following output (truncated):

```
                    Markov Switching Model Results
==========================================================================
Dep. Variable:                     y   No. Observations:              131
Model:             MarkovAutoregression   Log Likelihood          -181.263
Date:               Sun, 11 Jul 2021   AIC                        380.527
Time:                       18:36:04   BIC                        406.404
Sample:                   04-01-1951   HQIC                       391.042
                        - 10-01-1984
Covariance Type:              approx
                        Regime 0 parameters
==========================================================================
                coef    std err         z      P>|z|     [0.025     0.975]
--------------------------------------------------------------------------
const        -0.3588      0.265    -1.356      0.175     -0.877      0.160
                        Regime 1 parameters
==========================================================================
                coef    std err         z      P>|z|     [0.025     0.975]
--------------------------------------------------------------------------
const         1.1635      0.075    15.614      0.000      1.017      1.310
```

Figure 9.8: Markov Switching Model Results

We can see that we have two sets of parameters, one each for the two regimes. We also get measures of the statistical model quality (such as AIC and BIC).

At the bottom of the same output, we can see the regime transition parameters:

```
                    Regime transition parameters
==========================================================================
                coef    std err         z      P>|z|     [0.025     0.975]
--------------------------------------------------------------------------
p[0->0]       0.7547      0.097     7.819      0.000      0.565      0.944
p[1->0]       0.0959      0.038     2.542      0.011      0.022      0.170
==========================================================================
```

Figure 9.9: Regime transition parameters

These are the regime transitions we mentioned in the theory section on Markov Switching models.

Let's see the lengths of recession and expansion:

```
res_hamilton.expected_durations
```

The output array([4.07604793, 10.4258926]) is in financial quarters. Therefore, a recession is expected to take about four quarters (1 year) and an expansion 10 quarters (two and a half years).

Next, we'll plot the probability of recession at each point in time. However, this is more informative if we overlay indicators of recession by the National Bureau of Economic Research (NBER), which we can load up with pandas-dataloader:

```
from pandas_datareader.data import DataReader
from datetime import datetime
usrec = DataReader('USREC', 'fred', start=datetime(1947, 1, 1),
end=datetime(2013, 4, 1))
```

This gives us a DataFrame in which recessions are indicated. Here are the first five rows:

	USREC
DATE	
1947-01-01	0
1947-02-01	0
1947-03-01	0
1947-04-01	0
1947-05-01	0

Figure 9.10: Recession indicators by NBER

In the first five rows, there was no recession according to NBER indicators.

We'll now plot NBER recession indicators against the model regime predictions:

```
import matplotlib.pyplot as plt
_, ax = plt.subplots(1) ax.plot(res_hamilton.filtered_marginal_
probabilities[0]) ax.fill_between(
    usrec.index, 0, 1, where=usrec['USREC'].values,
    color='gray', alpha=0.3
)
ax.set(
    xlim=(dta_hamilton.index[4], dta_hamilton.index[-1]),
    ylim=(0, 1),
    title='Filtered probability of recession'
)
```

This gives us actual recession data against model predictions:

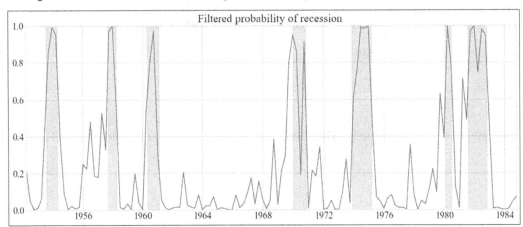

Figure 9.11: Filtered probability of recession

We can see there seems to be quite a good match between the model predictions and actual recession indicators.

Unfortunately, the `statsmodels` implementation doesn't provide the functionality for forecasting or out-of-sample prediction, so we'll end the brief demo here.

`Statsmodels` includes other datasets for regime-switching models to play around with.

In the following practice section, we'll apply Song and Chissom's model to a time-series forecasting problem using the `pyFTS` library for fuzzy time-series developed in the MINDS laboratory at the Universidade Federal de Minas Gerais (UFMG), Brazil.

Fuzzy Time-Series

In this section, we'll load two time-series of ticker symbols, the NASDAQ and the S&P 500 indices, over time and forecast them using Song and Chissom's 1993 algorithm. This follows closely the example tutorial in the library.

First, we'll install the library from the terminal (or similarly from the Anaconda navigator):

```
pip install pyFTS SimpSOM
```

Then, we'll define our datasets:

```
from pyFTS.data import NASDAQ, SP500
datasets = {
```

```
    "SP500": SP500.get_data()[11500:16000],
    "NASDAQ": NASDAQ.get_data()
}
```

Both datasets, the entries in our `datasets` dictionary, are vectors of roughly 4,000 scalar values. We'll take about 50% of these points for training, and we'll set this as a constant:

```
train_split = 2000
```

The model assumes a stationary process, so we'll need to preprocess our time-series by temporal differencing as discussed in *Chapter 2, Exploratory Time-Series Analysis with Time-Series.*

We'll define a first-order differencing operation for preprocessing:

```
from pyFTS.common import Transformations
tdiff = Transformations.Differential(1)
```

Let's plot our time-series and the transformation:

```
import matplotlib.pyplot as plt
fig, ax = plt.subplots(nrows=2, ncols=2)
for count, (dataset_name, dataset) in enumerate(datasets.items()):
  dataset_diff = tdiff.apply(dataset)
  ax[0][count].plot(dataset)
  ax[1][count].plot(dataset_diff)
  ax[0][count].set_title(dataset_name)
```

The plots of the original and transformed time-series look like this:

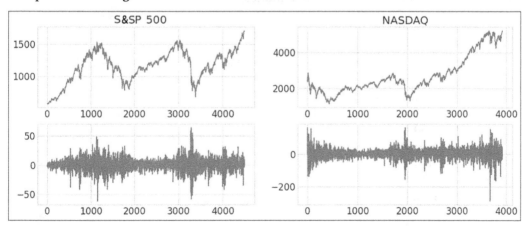

Figure 9.12: NASDAQ and S&P 500 — the original and transformed time-series

In the GitHub repository for this, you can see the Augmented Dickey-Fuller unit root test applied to the transformed time-series. This test for stationarity gives us the green light, and we continue with our model.

The next step is training our models for the two transformed (differenced) time-series:

```python
from pyFTS.models import song
from pyFTS.partitioners import Grid
models = {}
for count, (dataset_name, dataset) in enumerate(datasets.items()):
  partitioner_diff = Grid.GridPartitioner(data=dataset, npart=15,
transformation=tdiff)
  model = song.ConventionalFTS(partitioner=partitioner_diff)
  model.name = dataset_name
  model.append_transformation(tdiff)
  model.fit(
    dataset[:train_split],
    order=1
  )
  models[dataset_name] = model
```

We iterate over the datasets and train a separate model for each, which we save into a dictionary, `models`. The training consists of extracting relations from the training set.

As part of the model training, the preprocessed time-series is quantized as discussed in the theory section on fuzzy time-series models of this chapter.

We can plot our forecasts from the two models:

```python
_, ax = plt.subplots(nrows=2, ncols=1, figsize=[12, 6])

for count, (dataset_name, dataset) in enumerate(datasets.items()):
    ax[count].plot(dataset[train_split:train_split+200])
    model = models[dataset_name]
    forecasts = model.predict(dataset[train_split:train_split+200],
steps_ahead=1)
    ax[count].plot(forecasts)
    ax[count].set_title(dataset_name)

plt.tight_layout()
```

Again, we iterate over the two datasets. This time, we plot the original values in the test set (200 points) against estimated values predicted one step ahead. Please note that the models are not updated based on new data during the prediction.

This is our plot comparing the forecasts against the actual values in the test set:

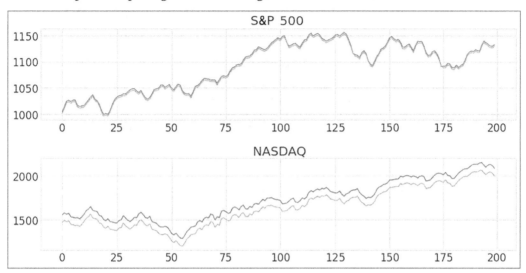

Figure 9.13: Fuzzy time-series forecast versus actual (S&P 500, NASDAQ).

Looking at these charts, the predictions look quite promising, but let's look at some hard numbers!

PyFTS has a convenience function to extract RMSE, MAPE, and finally, Theil's U, a measure of correlation. We introduced these measures in *Chapter 2, Exploratory Time-Series Analysis with Time-Series*.

```
from pyFTS.benchmarks import Measures
rows = []

for count, (dataset_name, dataset) in enumerate(datasets.items()):
    row = [dataset_name]
    test = dataset[train_split:train_split+200]
    model = models[dataset_name]
    row.extend(Measures.get_point_statistics(test, model))
    rows.append(row)

pd.DataFrame(
    rows,columns=["Dataset", "RMSE", "MAPE", "Theil's U"]
).set_index("Dataset")
```

We get these statistics:

	RMSE	MAPE	Theil's U
Dataset			
SP500	6.76	0.52	1.22
NASDAQ	90.85	5.14	3.79

Figure 9.14: Model statistics for fuzzy time-series modeling of NASDAQ and S&P 500

I'll leave it as an exercise for the reader to compare these two models with others based on these error metrics.

Bayesian Structural Time-Series Modeling

In this example, we'll apply BSTS modeling to understand the causal effect of treatment in time-series.

First, we'll install the library:

```
pip install tfcausalimpact
```

Now, we'll load a dataset, and we'll estimate the consequence of a treatment.

Here, we'll be estimating the impact of the emissions scandal of September 2015 for Volkswagen. We'll work with the stock values of three big companies, Volkswagen, BMW, and Allianz. The dataset comes with the Python Causal Impact (tfcausalimpact) library:

```
import pandas as pd
from causalimpact import CausalImpact
data = pd.read_csv("https://raw.githubusercontent.com/WillianFuks/
tfcausalimpact/master/tests/fixtures/volks_data.csv", header=0, sep=' ',
index_col='Date', parse_dates=True)
```

Now we have the stock values. Let's plot them:

```
data.plot()
```

Here are the stocks over time:

Figure 9.15: Stock values of three big companies (Volkswagen, BMW, Allianz)

We can see a sharp drop in the value of Volkswagen shares in late 2015. Let's try to find out the actual impact of the emission scandal. We can build our model like this:

```
pre_period = [str(np.min(data.index.values)), "2015-09-13"]
post_period = ["2015-09-20", str(np.max(data.index.values))]
ci = CausalImpact(data.iloc[:, 0], pre_period, post_period, model_
args={'nseasons': 52, 'fit_method': 'vi'})
```

The model statistics provide us with the causal impact estimate:

```
print(ci.summary())
```

We see these stats here:

```
Posterior Inference {Causal Impact}
                          Average              Cumulative
Actual                    126.91               10026.07
Prediction (s.d.)         171.28 (17.33)       13531.49 (1369.17)
95% CI                    [136.07, 204.01]     [10749.78, 16116.83]
Absolute effect (s.d.)    -44.37 (17.33)       -3505.42 (1369.17)
95% CI                    [-77.1, -9.16]       [-6090.76, -723.71]
Relative effect (s.d.)    -25.91% (10.12%)     -25.91% (10.12%)
95% CI                    [-45.01%, -5.35%]    [-45.01%, -5.35%]


Posterior tail-area probability p: 0.01
Posterior probability of a causal effect: 99.2%
```

Figure 9.16: Causal impact estimates and model statistics

As discussed before, the Causal Impact model developed by Google works by fitting a BSTS model to observed data, which is later used to predict what the results would be had no intervention happened in a given time period.

The total estimated effect is about 44 points — the stock price would be 44 points higher if not for the emissions scandal. The impact summary report gives us this analysis (excerpt):

```
During the post-intervention period, the response variable had
an average value of approx. 126.91. By contrast, in the absence of an
intervention, we would have expected an average response of 171.28. The
95% interval of this counterfactual prediction is [136.07, 204.01].
Subtracting this prediction from the observed response yields
an estimate of the causal effect the intervention had on the
response variable. This effect is -44.37 with a 95% interval of [-77.1,
-9.16]. For a discussion of the significance of this effect, see below.
```

Figure 9.17: Causal impact analysis report

This gives us a very good idea of what the model estimates.

We can plot the effect as well:

```
ci.plot(panels=["original"]
```

The plot is as follows:

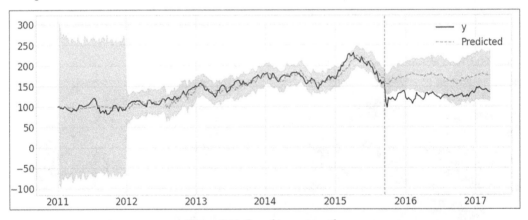

Figure 9.18: Causal impact graph

Again, we see the original time-series against the predicted counterfactual value.

The emissions scandal wiped out a massive amount of value from Volkswagen. The 44 points can give us a monetary value of how much cheating emissions tests cost Volkswagen.

Summary

In this chapter, we've discussed how probabilistic models for time-series can help us make decisions with an estimate of uncertainty in the context of financial forecasting. These forecasts drive business decisions for financial planning.

I've introduced Prophet, Markov models, and Fuzzy time-series models. We've discussed the components of Facebook's Prophet model. For Markov models, we've discussed the main ideas, such as the Markov property, and we've discussed more details regarding Switching Models. Then I've explained some basics of fuzzy set theory and how this is applied to time-series.

Finally, we've delved into the intuition and some of the theory of BSTS models in the context of estimating treatment effects in experiments.

Finally, we went through an applied exercise with each method. In the BSTS practice, we've looked at the effect of the Volkswagen emissions scandal.

10

Deep Learning for Time-Series

Deep learning is a subfield of machine learning concerned with algorithms relating to neural networks. Neural networks, or, more precisely, **artificial neural networks (ANNs)** got their name because of the loose association with biological neural networks in the human brain.

In recent years, deep learning has been enhancing the state of the art across the bench in many application domains. This is true for unstructured datasets such as text, images, video, and audio; however, tabular datasets and time-series have so far shown themselves to be less amenable to deep learning.

Deep learning brings a very high level of flexibility and can offer advantages of both online learning, as discussed in *Chapter 8, Online Learning for Time-Series*, and probabilistic approaches, as discussed in *Chapter 9, Probabilistic Models for Time-Series*. However, with its highly parameterized models, finding the right model can be a challenge.

Among the contributions deep learning has been able to bring to time-series are data augmentation, transfer learning, long sequence time-series forecasts, and data generation with **generative adversarial networks (GANs)**. However, it's only very recently that deep learning approaches have become competitive in relation to forecasting, classification, and regression tasks.

In this chapter, we'll discuss deep learning applied to time-series, looking, in particular, at algorithms and approaches designed for time-series. We'll get into current challenges, promising avenues of research, and competitive approaches that bring deep learning to time-series. We'll go into detail about a lot of the recent innovations in deep learning for time-series.

We're going to cover the following topics:

- Introduction to deep learning
- Deep learning for time-series
  - Autoencoders
  - InceptionTime
  - DeepAR
  - N-BEATS
  - Recurrent neural networks
  - ConvNets
  - Transformer architectures
  - Informer
- Python practice
  - Fully connected network
  - Recurrent neural network
  - Dilated causal convolutional neural network

Let's start with an introduction to deep learning and the core concepts.

Introduction to deep learning

Deep learning is based on fundamental concepts that find their roots early in the 20th century – the wiring between neurons. Neurons communicate chemically and electrically through so-called neurites.

This wiring was first described and drawn by Santiago Ramón y Cajal, a Spanish neuroscientist. He charted the anatomy of the brain and the structure of neural networks in the brain. He received the Nobel Prize in Physiology or Medicine in 1906, which he shared with Camillo Golgi, who invented the stains for neurons based on potassium dichromate and silver nitrate that Ramón y Cajal applied in his microscopy studies.

The chart below is just one of his elaborate drawings of the arborization of neural connections (called neurites – dendrites and axons) between neurons in the brain (source Wikimedia Commons):

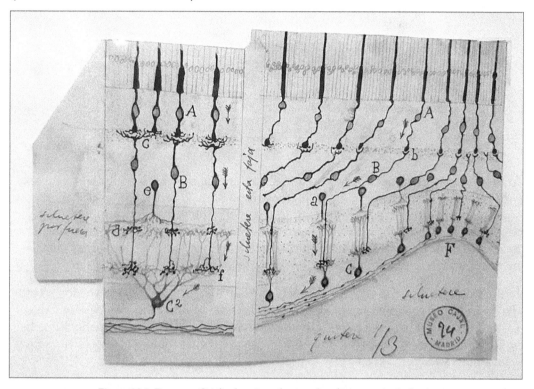

Figure 10.1: Ramon y Cajal's drawing of networks of neurons in the brain

In the schematic, you can appreciate neurons as gray dots in layers of the brain. Between neurons are dendrites and axons, the wiring of the brain. Each neuron takes up what amounts to information about the environment through gate stations to neurites that are called synapses.

Ramón y Cajal and his pupils brought to life *cable theory*, where the electric current passing through neurites is modeled by mathematical models. The voltage arriving at neural sites through the dendrites that receive synaptic inputs at different sites and times was recognized as sensory and other information was transmitted between cells. This is the foundation of today's detailed neuron models employed in research to model synaptic and neural responses.

The basic function of neurons was formalized by Frank Rosenblatt in 1958 as the perceptron – a model that contains the essentials of most modern deep learning concepts.

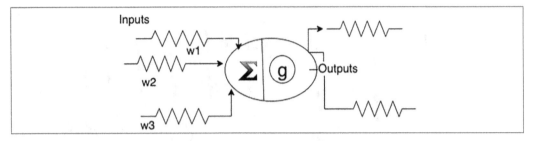

Figure 10.2: The perceptron model

In the perceptron model, a neuron – illustrated by the oval in the middle – receives input from other neurons. In a model, these inputs could represent text, images, sounds, or any other type of information. These get integrated by summing them up. In this sum, each input from a neuron, i, comes with its weight, w_i, that marks its importance. This integrated input can then lead to a neural activation as given by the neuron's activation function, g.

In the simplest case, the activation function could just be a threshold function so that the neuron gets activated if the weighted sum of the inputs exceeds a certain value. In modern neural networks, the activation functions are non-linear functions, such as sigmoid functions or the rectified linear function where the output is linear above a threshold and cropped below.

When the network is stimulated by data, input neurons are activated and feed second-order neurons, which then feed other neurons in turn until the output layers are activated. This is called *feedforward propagation*.

The perceptron consisted of a single layer of integrating neurons that sum input over their incoming connections. It was demonstrated by Marvin Minsky and Seymour Pappert in their book *Perceptrons* (1969) that these neurons, similar to a simple linear model, cannot approximate complex functions that are relevant in the real world.

Multilayer neural networks can overcome this limitation, however, and this is where we slowly enter into the realm of deep learning. These networks can be trained through an algorithm called *backpropagation* – often credited to Paul Werbos (1975). In backpropagation, outputs can be compared to targets and the error derivative can be fed back through the network to calculate adjustments to the weights in the connections.

Another innovation in neural networks again comes from neuroscience. In the 1950s and 1960s, David Hubel and Torsten Wiesel found that neurons in the cat visual cortex (V1) respond to small regions of the visual field. This region they termed the receptive field (1959, "Receptive fields of single neurons in the cat's striate cortex"). They distinguished between two basic cell types:

- Simple cells – these cells can be characterized largely by a summation over the inputs
- Complex cells – cells that respond to a variety of stimuli across different locations

Complex cells have inspired computational layers in neural networks that employ convolutions, first by Kunihiko Fukushima in 1980. We've discussed convolutions in *Chapter 3*, *Preprocessing Time-Series*.

Neural networks with convolutional layers are the predominant type of model for applications such as image processing, classification, and segmentation. Yann LeCun and colleagues introduced the LeNet architecture (1989), where convolution kernels are learned through backpropagation for the classification of images of hand-written numbers.

Deep learning networks often come not just with layers, where inputs get propagated from one layer to the next (feedforward). The connections can also be recurrent, where they connect to the neurons of the same layer or even back to the same neuron.

A recurrent neural network framework, **long short-term memory (LSTM)** was proposed by Jürgen Schmidhuber and Sepp Hochreiter in 1997. LSTMs can retrieve and learn information for a longer period of time compared to previous models. This model architecture was, for some time, powering industry models such as speech recognition software for Android smartphones, but has since been mostly replaced by convolutional models.

In 2012, AlexNet, created by Alex Krizhevsky in collaboration with Ilya Sutskever and Geoffrey Hinton, made a breakthrough in the ImageNet Large Scale Visual Recognition Challenge (short ImageNet), where millions of images are to be categorized between 20,000 categories. AlexNet brought down the top-5 error rate from around 25% to about 15%. The model, utilizing massively parallel hardware powered by **Graphics Processing Units (GPUs)**, combined fully connected layers with convolutions and pooling layers.

This was only the beginning of a radical performance improvement on different tasks, including images. The AlexNet performance was beaten the following year by ResNet.

Since the ResNet paper is highly influential, it's worth taking a short detour to explain how it works. ResNets were introduced by Kaiming He and others at Microsoft Research in 2015 ("*Deep Residual Learning for Image Recognition*"). A common problem with deep neuron networks is that their performance can saturate and degrade with more layers added partly because of the vanishing gradient problem, where the error gradient calculated in the optimization will become too small to be useful.

Inspired by pyramidal cells in the brain, residual neural networks employ so-called *skip connections*, essentially shortcuts to jump intermediate layers. A ResNet is a network that contains blocks with skip connections (residual blocks), as indicated in this schema:

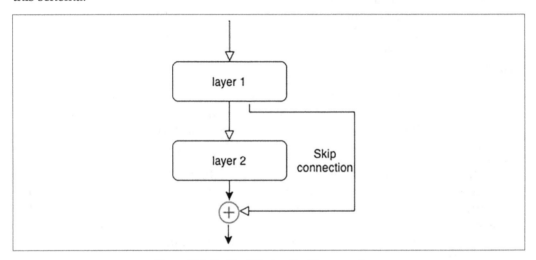

Figure 10.3: Residual block with skip connections

In the residual block illustrated, the output of layer 2 is as follows:

$$H(x) = G_{layer\ 2}(x) + G_{skip}(x),$$

where $G_{layer\ 2}$ and G_{skip} are the activation functions in layer 2 and the skip connections, respectively. G_{skip} is often the identify function, where activations of layer 1 are unchanged. If the dimensionality between layer 1 and layer 2 outputs don't match, either padding or convolutions are used.

With these skip connections, Kaiming He and others successfully trained networks with as many as 1,000 layers. The original ResNet from 2015 was very successful on images. Among other accolades it collected was winning several top competitions for image classification and object detection: ILSVRC 2015, ILSVRC 2015, COCO 2015 competition in ImageNet Detection, ImageNet localization, Coco detection, and Coco segmentation.

I've summarized the early history of ANNs and deep learning in the timeline here:

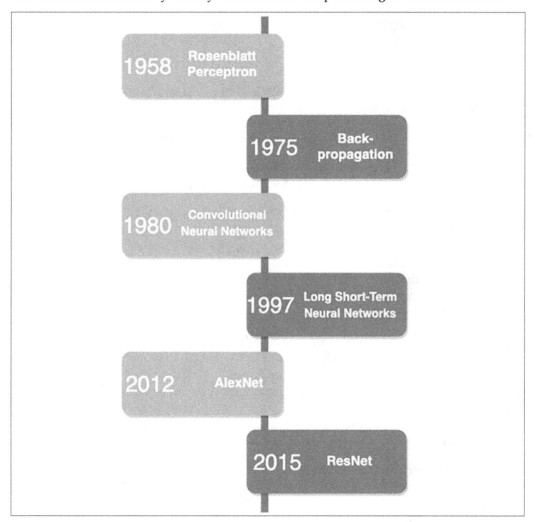

Figure 10.4: Timeline of artificial neural networks and deep learning

Please note that this is highly simplified, leaving out many important milestones. I've ended in 2015, when ResNet was presented.

There are a host of architectures and approaches in deep learning, and this chart displays a typology of these methodologies:

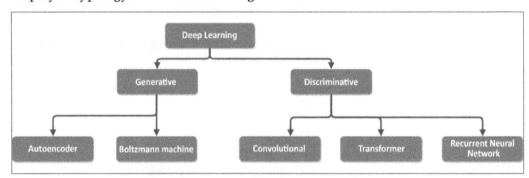

Figure 10.5: Typology of deep learning approaches

We've mentioned a few of these approaches in this section, and we'll explain a few of these methods in more detail in the next sections as they are relevant to time-series.

The computational complexity of techniques based on deep neural networks is driven in the first instance by the dimension of the input data and depends on the number of hidden layers trained using backpropagation. High-dimensional data tend to require more hidden layers to ensure a higher hierarchy of feature learning, where each layer derives higher-level features based on the previous level. The training time and complexity increase with the number of neurons – the number of hyperparameters can sometimes reach the millions or billions.

The representational power of deep learning that constructs a stack of derived features as part of the learning allows modelers to get away from hand-crafted features. Further advantages of using deep learning models include their flexibility in terms of choosing architecture, hyperparameters such as activation functions, regularization, layer sizes, and loss objectives, but this is traded off against their complexity in terms of the number of parameters, and the difficulty of interrogating their inner workings.

Deep learning methods offer better representation and, in consequence, prediction on a multitude of time-series datasets compared to other machine learning approaches; however, they haven't found the impact so far that they had in other areas.

Deep learning for time-series

Recent years have seen a proliferation of deep neural networks, with unprecedented improvements across various application domains, in particular images, natural language processing, and sound. The potential advantage of deep learning models is that they can be much more accurate than other types of models, thereby pushing the envelope in domains such as vision, sound, and **natural language processing (NLP)**.

In forecasting, especially demand forecasting, data is often highly erratic, discontinuous, or bursty, which violates the core assumptions of classical techniques, such as Gaussian errors, stationarity, or homoscedasticity, as discussed in *Chapter 5, Forecasting of Time-Series*. Deep learning techniques applied to forecasting, classification, or regression tasks could overcome many of the challenges faced by classical approaches, and, most importantly, they could provide a way to model non-linear dynamics usually neglected by traditional methods such as Box-Jenkins, Exponential Smoothing (ES), or state-space models.

Many deep learning algorithms have been applied more recently to time-series, both with univariate and multivariate time-series. The model architectures encompass recurrent neural networks (RNNs), most prominently long short-term memory (LSTM) models, and transformer and convolutional models, or different types of autoencoders.

As regards their application to time-series, however, they haven't been able to challenge the top models in the field. For instance, as pointed out by Spyros Makridakis and others (2020), in the M4 competition, arguably the most important benchmark for univariate time-series forecasting, the best-ranking methods were ensembles of widely used classical statistical techniques rather than pure machine learning methods.

This could have been at least partly due to the nature of the competition. As pointed out by Slawek Smyl, de-seasonalization of seasonal series was very important in the M4 competition, given that the series were provided as scalar vectors without timestamps, so there was no way to incorporate calendar features such as the day of the week or the month number.

In the M4 competition, out of 60 competition entries, the first machine learning method ranked at place 23. However, interestingly, the winner of the M4 competition was a hybrid between a dilated LSTM with attention and a Holt-Winters statistical model. Another top contender, developed by the research group around Rob Hyndman, applied a gradient boosted tree ensemble to outputs from traditional models (*FFORMA: Feature-based Forecast Model Averaging*, 2020).

These rankings led Spyros Makridakis and others to conclude that hybrids or mixtures of classical and machine learning methods are the way forward. The search is ongoing for a deep learning architecture that could provide an inflection point in research and applications similar to that of AlexNet or Inception for the image domain.

In *Chapter 4, Introduction to Machine Learning with Time-Series*, we discussed first how difficult it is to beat baseline approaches such as Nearest Neighbor with **Dynamic Time Warping (DTW)** and then state-of-the-art approaches. The most competitive model in terms of performance is **HIVE-COTE (Hierarchical Vote Collective of Transformation-Based Ensembles)**, which consists of ensembles of machine learning models – very expensive in terms of resources, owing to the number of computations and the long runtime.

The sardonic reader might comment that this sounds like deep learning already and ask whether deep learning hasn't already taken over as the state-of-the-art method. Generally speaking, the complexity of deep learning models is much higher than that of traditional models or other machine learning techniques. The case can be made that this is one of the biggest distinguishing characteristics of deep learning models.

Is there a deep learning model architecture of similar or lower complexity than HIVE that can achieve competitive results?

I've summarized a few libraries that implement algorithms with deep learning for time-series in this table:

Library	Maintainer	Algorithms	Framework
dl-4-tsc	Hassan Ismail Fawaz	**Multi-Layer Perceptron (MLP)**, **Fully Connected Network (FCN)**, ResNet, Encoder (based on CNN), **Multi-Scale Convolutional Neural Network (MCNN)**, **Time Le-Net (t-LeNet)**, **Multi-Channel Deep Convolutional Neural Network (MCDCNN)**, Time-CNN, **Time Warping Invariant Echo State Network (TWIESN)**, InceptionTime	TensorFlow/ Keras

Sktime-DL	Students and staff at the University of East Anglia around Tony Bagnell	ResNet, CNN, InceptionTime (through an interface with another library)	TensorFlow/Keras
Gluon-TS	Amazon Web Services – Labs	Gluon-TS specializes in probabilistic neural network models such as these: **Convolutional Neural Network (CNN)**, DeepAR, **Recurrent Neural Network (RNN)**, **Multi-Layer Perceptron (MLP)**	MXNET
Pytorch Forecasting	Daniel Hoyos and others	Recurrent Networks (GRU, LSTM), Temporal Fusion Transformers, N-Beats, Multilayer Perceptron, DeepAR	PyTorch Lightning

Figure 10.6: Overview of several deep learning libraries for time-series

Sktime-DL is an extension to sktime, maintained by the same research group. As of August 2021, this library is undergoing a rewrite.

Gluon-TS is based on the MXNET deep learning modeling framework, and – apart from the network architectures noted in the table – includes many other features, such as kernels for **Support Vector Machines** (**SVMs**) and **Gaussian Process** (**GP**), and distributions for probabilistic network models.

dl-4-tsc is the GitHub companion repository for a review paper of many time-series deep learning algorithms, prepared by Hassan Ismail Fawaz and others (2019). It includes implementations in TensorFlow/Keras of their implementations. It is not a library *per se*, as it isn't installed like a library and the models run with datasets; however, since the algorithms are implemented in TensorFlow and Keras, anyone with a knowledge of these will feel at home.

Pytorch-forecasting, sktime-DL, and Gluon-TS come with their own abstractions of datasets that help with the automation of common tasks. While Sktime-DL builds on the sktime abstractions, Pytorch-Forecasting and Gluon-TS have batteries built-in for deep learning with utilities for common tasks such as the scaling and encoding of variables, normalizing the target variable, and downsampling. These abstractions come at the cost of a learning curve, however, and I should caution the impatient reader that it can take time to get up to speed with this, which is why I am omitting them from the practice part.

I've omitted repositories from this table that only implement a single algorithm. In the next visualization, I have included some of these, such as the repositories for the Informer model or the neural prophet. In the following diagram, you can see the popularity of a few repositories for time-series deep learning:

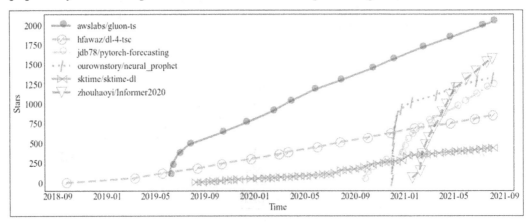

Figure 10.7: Popularity of deep learning libraries for time-series

As always, I've tried to choose the most popular repositories – and repositories that have been updated recently. You can see that Gluon-TS is the most popular repository. Of the repositories implementing several algorithms, Pytorch Forecasting comes closest and has been making inroads recently in terms of popularity.

In the next sections, we'll concentrate on recent and competitive approaches with deep learning on time-series. We'll go through a few of the most prominent algorithms in more detail: Autoencoders, InceptionTime, DeepAR, N-BEATS, RNNs (most prominently LSTMs), ConvNets, and Transformers (including the Informer).

Autoencoders

Autoencoders (AEs) are artificial neural networks that learn to efficiently compress and encode data and are trained on reconstruction errors. A basic linear AE is essentially functionally equivalent to a **Principal Component Analysis (PCA)**, although, in practice, AEs are often regularized.

AEs consist of two parts, the encoder and the decoder, as illustrated below (source: Wikipedia):

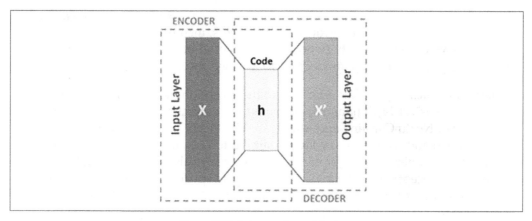

Figure 10.8: Autoencoder architecture

Encoders and decoders are often both of the same architecture, which depends on the domain. For example, as regards images, they often contain convolutions like LeNet. For modeling time dependence, they can include causal convolutions or recurrent layers as a way of modeling time dependence.

AEs are a natural way of reducing noise. They are often employed for anomaly detection in time-series.

InceptionTime

In a massive test that took about a month to complete on a cluster of 60 GPUs, Hassan Ismail Fawaz and others at the Université de Haute Alsace ran deep learning algorithms on the univariate UCR/UEA time-series classification archive (85 time-series) and the 13 datasets of the **Multivariate Time-Series (MTS)** classification archive. They presented this work in their paper *"Deep learning for time-series classification: a review"* in 2019.

They conducted a systematic evaluation of 11 models, including LeNet, **Fully Connected Networks (FCNs)**, Time-CNN, and ResNet. Only 9 of the algorithms completed all the tests. Compared to deep learning algorithms on the univariate datasets (UCR/UEA), ResNet won on 50 problems out of 85, and it was statistically better than the next best algorithm, the **fully convolutional neural network (FCNN)**. At the same time, it was not statistically worse than HIVE-COTE, the top model in time-series classification. On the multivariate benchmark, the FCNN won, although they couldn't find any statistically significant differences between networks.

In another paper, *"InceptionTime: Finding AlexNet for Time-Series Classification,"* Hassan Ismail Fawaz and an extended group of researchers, including Geoff Webb and others from Monash University (who we encountered in *Chapter 3, Preprocessing Time-Series*), presented a new model that they called InceptionTime.

The name *InceptionTime* makes reference to the Inception model (*"Going Deeper with Convolutions"*, 2014), a network presented by researchers at Google, and the universities of North Carolina and Michigan. The Inception architecture consists of feedforward and convolutional layers, similar to LeNet, which we mentioned earlier in this chapter. A 22-layer variant was therefore also called GoogleLetNet (alternatively: Inception model version 1). Roughly speaking, the inception model consists of modules ("inception modules") that concatenate convolutions of different sizes together.

InceptionTime takes ensembles of inception-type models with different hyperparameters (filters of varying lengths). They experimented with the number of networks in the ensemble and the filter sizes and finally showed that their model significantly outperformed ResNet on the 85 datasets of the UCR archive, while being statistically on a par with HIVE-COTE – with a much-reduced training time compared to HIVE-COTE.

DeepAR

DeepAR is a probabilistic forecasting method coming out of Amazon Research Germany. It is based on training an auto-regressive recurrent network model. In their article *"DeepAR: Probabilistic forecasting with autoregressive recurrent networks"* (2019), David Salinas, Valentin Flunkert, and Jan Gasthaus demonstrated through extensive empirical evaluation on several real-world forecasting datasets (parts, electricity, traffic, ec-sub, and ec) accuracy improvements of around 15% compared to state-of-the-art methods.

DeepAR was designed for demand forecasting and consists of an RNN architecture and incorporates a negative binomial likelihood for unbounded count data that can stretch across several orders of magnitude. Monte Carlo sampling is used to compute quantile estimates for all sub-ranges in the prediction horizon. For the case when the magnitudes of the time-series vary widely, they also introduced a scaling of the mean and variance parameters of the negative binomial likelihood by factors that depend on the mean of the time-series and the output of the network.

N-BEATS

This is a model architecture for univariate times series point forecasting based on backward and forward residual links and a very deep multilayer network of fully connected ReLU neurons. N-BEATS uses deep learning primitives such as residual blocks instead of any time-series-specific components and is the first architecture to demonstrate that deep learning using no time-series-specific components can outperform well-established statistical approaches.

Published in 2019 by a group around Yoshua Bengio (*"N-BEATS: Neural basis expansion analysis for interpretable time-series forecasting"*), this network reached state-of-the-art performance for two configurations and outperformed all other methods, including ensembles of traditional statistical methods in benchmarks over the M3, M4, and TOURISM competition datasets.

A common criticism of deep learning is the opaque nature of the learning or – inversely – a lack of transparency in terms of what the network does. N-BEATS can be made interpretable with few changes without losing significant accuracy.

Recurrent neural networks

RNNs, most prominently LSTMs, have been applied a lot to multivariate electricity consumption forecasting. Electricity forecasting is a long sequence time-series, where it's necessary to precisely capture the long-range correlation coupling between items of a sequence over time.

Earlier works explored combinations of LSTMs with dilation, residual connections, and attention. These served as a basis for the winner of the M4 competition (Slawek Smyl, 2020).

Smyl introduced a mixture of a standard **Exponential Smoothing** (**ES**) model with LSTM networks. The ES equations enable the method to effectively capture the main components of the individual series, such as seasonality and baseline level, while the LSTM networks can model the nonlinear trend.

A problem with RNNs, including LSTMs, is that they cannot be parallelized easily at the cost of training time and computing resources. It has also been argued that LSTMs cannot capture long-range dependencies since they struggle with sequences longer than about 100 time steps. RNNs encode past hidden states to capture dependencies with previous items, and they show a decrease in performance due to long dependencies.

In the next section, we'll look at the transformer architecture, which has been taking over from LSTM models both in terms of performance and, more recently, in popularity.

It has been shown that convolutional architectures can outperform recurrent networks on tasks such as audio processing and machine translation, and they have been applied to time-series tasks as well.

ConvNets

Researchers from Carnegie Mellon University and Intel Labs ("*An Empirical Evaluation of Generic Convolutional and Recurrent Networks for Sequence Modeling*" 2018) compared generic convolutional and recurrent networks such as LSTM/GRU for sequence modeling across a broad range of tasks. These were large textual datasets and posed sequence problems, such as the problem of addition, the copying of memory tasks, or polyphonic music. Problems and datasets such as these are commonly used to benchmark recurrent networks.

They found that a simple convolutional architecture, the **Temporal Convolutional Network** (**TCN**), performs better than RNNs on a vast range of tasks' canonical recurrent networks (such as LSTMs) while demonstrating a longer effective memory.

The important characteristic of the convolutions in the TCNs is that they are causal. A convolution is causal if its output is the result of only current and past inputs. This is illustrated here (source: keras-tcn, GitHub):

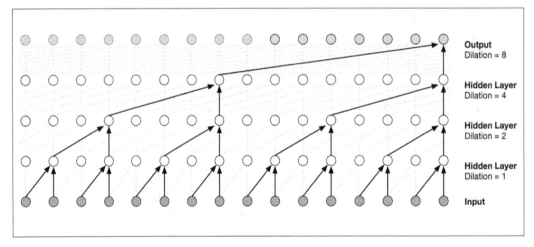

Figure 10.9: A time-causal convolution

An output at time t is convolved only with elements from time t and earlier. This means that information from the future cannot leak to the past. A disadvantage of this basic design is that in order to achieve a long effective history size, we need an extremely deep network or very large filters.

Some advantages of convolutions over RNNs are parallelism, the flexible receptive field size (specifying how far the model can see), and stable gradients – backpropagation through time comes with the vanishing gradient problem.

The transformer also addresses the perceived shortcomings of RNNs.

Transformer architectures

Transformers, introduced in the article "*Attention is all you need*" (2017) by researchers at Google Brain and the University of Toronto, were designed to avoid recursion in order to allow parallel computation.

Transformers introduced two building blocks – multi-head attention and positional embeddings. Rather than working sequentially, sequences are processed as a whole rather than item by item. They employ self-attention, where similarity scores between items in a sentence are stored.

Transformers were introduced originally for machine translation, where they were shown to outperform Google's Neural Machine Translation models. The central piece is therefore the alignment of two sequences. Instead of recurrence, positional embeddings were introduced where weights encode information related to a specific position of a token in a sequence.

Transformers consist of stacked modules, first encoder and then decoder modules. Encoder modules each consist of a self-attention layer and a feed-forward layer, while decoder modules consist of self-attention, encoder-decoder attention, and a feed-forward layer. These modules can be stacked, thereby creating very large models that can learn massive datasets.

Transformers have pushed the envelope in NLP, especially in translation and language understanding. Furthermore, OpenAI's powerful GPT-3 model for language generation is a transformer as well, as is DeepMind's AlphaFold 2, a model that predicts protein structure from their genetic sequences.

Transformers have been able to maintain performance across longer sequences. However, they can capture only dependencies within the fixed input size used to train them. To work with even longer sentences beyond the fixed input width, architectures such as Transformer-XL reintroduce recursion by storing hidden states of already encoded sentences to leverage them in the subsequent encoding of the next sentences.

In the article "*Temporal Fusion Transformers for Interpretable Multi-horizon Time-Series Forecasting*," researchers from the University of Oxford and Google Cloud AI introduced an attention-based architecture, which they called a **Temporal Fusion Transformer** (**TFT**). To learn temporal relationships at different scales, TFT uses recurrent layers for local processing and interpretable self-attention layers for long-term dependencies. Furthermore, a series of gating layers suppress unnecessary components.

On a variety of real-world datasets, they demonstrated significant performance improvements of their architecture over a broad benchmark set. Among other improvements, they outperformed Amazon's DeepAR by between 36 and 69%.

Informer

A problem with transformers is the quadratic time complexity and memory usage, along with the limitations of the encoder-decoder architecture. This complicates predictions over longer time periods, for example, 510 time steps of hourly electricity consumption. To address these issues, researchers from Beihang University, UC Berkeley, Rutgers, and SEDD Company designed an efficient transformer-based model for long sequence time-series forecasting, named Informer – "*Informer: Beyond Efficient Transformer for Long Sequence Time-Series Forecasting*". The paper obtained the Outstanding Paper Award at the AAAI conference in 2021.

The generative decoder alleviates the time complexity of the encoder-decoder by predicting long time-series sequences in one forward operation rather than step-by-step. They replaced the positional embeddings with a new self-attention mechanism called ProbSparse Self-Attention, which achieves $O(L \log L)$, where L is the length of the sequence, instead of quadratic time complexity and memory usage, $O(L^2)$, while maintaining a comparable performance on sequence alignment.

Finally, the self-attention distillation halves the cascading layer input, and efficiently handles extreme long input sequences. This reduces the complexity from $O(J\ L2)$, where J is the number of transformer layers, to $O((2 - \varepsilon)L \log L)$.

The Informer architecture is illustrated in this schema (from the official Informer repository):

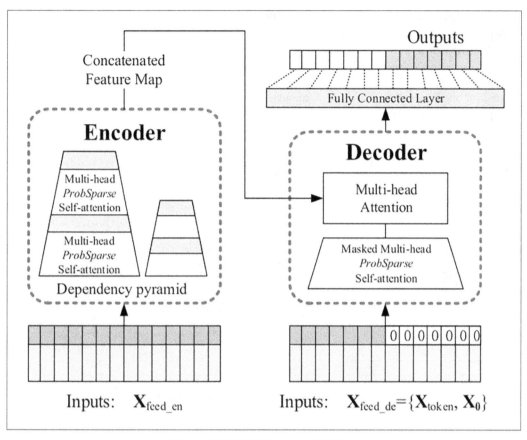

Figure 10.10: The Informer architecture

This diagram shows that the Informer significantly outperforms existing methods on datasets of long time-series forecasting such as Electricity Transformer Temperature (ETT), Electricity Consuming Load (ECL), and Weather.

On univariate datasets, they achieved superior performance compared with all competitors, except for two cases, where DeepAR was slightly better, as shown here (source: Informer GitHub repository):

Methods		Informer	Informer†	LogTrans	Reformer	LSTMa	DeepAR	ARIMA	Prophet
Metric		MSE MAE	MSE MAE	MSE MAE	MSE MAE	MSE MAE	MSE MAE	MSE MAE	MSE MAE
ETTh1	24	0.098 0.247	**0.092 0.246**	0.103 0.259	0.222 0.389	0.114 0.272	0.107 0.280	0.108 0.284	0.115 0.275
	48	**0.158 0.319**	0.161 0.322	0.167 0.328	0.284 0.445	0.193 0.358	0.162 0.327	0.175 0.424	0.168 0.330
	168	**0.183 0.346**	0.187 0.355	0.207 0.375	1.522 1.191	0.236 0.392	0.239 0.422	0.396 0.504	1.224 0.763
	336	0.222 0.387	**0.215 0.369**	0.230 0.398	1.860 1.124	0.590 0.698	0.445 0.552	0.468 0.593	1.549 1.820
	720	0.269 0.435	**0.257 0.421**	0.273 0.463	2.112 1.436	0.683 0.768	0.658 0.707	0.659 0.766	2.735 3.253
ETTh2	24	**0.093 0.240**	0.099 0.241	0.102 0.255	0.263 0.437	0.155 0.307	0.098 0.263	3.554 0.445	0.199 0.381
	48	**0.155 0.314**	0.159 0.317	0.169 0.348	0.458 0.545	0.190 0.348	0.163 0.341	3.190 0.474	0.304 0.462
	168	**0.232 0.389**	0.235 0.390	0.246 0.422	1.029 0.879	0.385 0.514	0.255 0.414	2.800 0.595	2.145 1.068
	336	0.263 **0.417**	**0.258** 0.423	0.267 0.437	1.668 1.228	0.558 0.606	0.604 0.607	2.753 0.738	2.096 2.543
	720	**0.277 0.431**	0.285 0.442	0.303 0.493	2.030 1.721	0.640 0.681	0.429 0.580	2.878 1.044	3.355 4.664
ETTm1	24	**0.030 0.137**	0.034 0.160	0.065 0.202	0.095 0.228	0.121 0.233	0.091 0.243	0.090 0.206	0.120 0.290
	48	0.069 0.203	**0.066 0.194**	0.078 0.220	0.249 0.390	0.305 0.411	0.219 0.362	0.179 0.306	0.133 0.305
	96	0.194 **0.372**	**0.187** 0.384	0.199 0.386	0.920 0.767	0.287 0.420	0.364 0.496	0.272 0.399	0.194 0.396
	288	**0.401 0.554**	0.409 0.548	0.411 0.572	1.108 1.245	0.524 0.584	0.948 0.795	0.462 0.558	0.452 0.574
	672	**0.512 0.644**	0.519 0.665	0.598 0.702	1.793 1.528	1.064 0.873	2.437 1.352	0.639 0.697	2.747 1.174
Weather	24	**0.117 0.251**	0.119 0.256	0.136 0.279	0.231 0.401	0.131 0.254	0.128 0.274	0.219 0.355	0.302 0.433
	48	**0.178** 0.318	0.185 **0.316**	0.206 0.356	0.328 0.423	0.190 0.334	0.203 0.353	0.273 0.409	0.445 0.536
	168	**0.266 0.398**	0.269 0.404	0.309 0.439	0.654 0.634	0.341 0.448	0.293 0.451	0.503 0.599	2.441 1.142
	336	**0.297 0.416**	0.310 0.422	0.359 0.484	1.792 1.093	0.456 0.554	0.585 0.644	0.728 0.730	1.987 2.468
	720	**0.359 0.466**	0.361 0.471	0.388 0.499	2.087 1.534	0.866 0.809	0.499 0.596	1.062 0.943	3.859 1.144
ECL	48	0.239 0.359	0.238 0.368	0.280 0.429	0.971 0.884	0.493 0.539	**0.204 0.357**	0.879 0.764	0.524 0.595
	168	0.447 0.503	0.442 0.514	0.454 0.529	1.671 1.587	0.723 0.655	**0.315 0.436**	1.032 0.833	2.725 1.273
	336	0.489 0.528	0.501 0.552	0.514 0.563	3.528 2.196	1.212 0.898	**0.414 0.519**	1.136 0.876	2.246 3.077
	720	**0.540 0.571**	0.543 0.578	0.558 0.609	4.891 4.047	1.511 0.966	0.563 0.595	1.251 0.933	4.243 1.415
	960	**0.582 0.608**	0.594 0.638	0.624 0.645	7.019 5.105	1.545 1.006	0.657 0.683	1.370 0.982	6.901 4.264
Count		32	12	0	0	0	6	0	0

Figure 10.11: Univariate long sequence time-series forecasting performance

Most significantly, they beat competitors such as ARIMA, prophet, LSTMs, and other transformer-based architectures.

On a multivariate benchmark, they also beat competitors including other transformer-based models and LSTMs.

We'll put some of this into practice now.

Python practice

Let's model airplane passengers. We'll forecast the monthly number of passengers.

This dataset is considered one of the classic time-series, published by George E.P. Box and Gwilym Jenkins alongside the book "Time-Series Analysis: Forecasting and Control" (1976). I have provided a copy of this dataset in the `chapter10` folder of the book's GitHub repository. You can download it from there or use the URL directly in `pd.read_csv()`.

We'll first start with a simple FCN and then we'll apply a recurrent network, and finally, we'll apply a very recent architecture, a Dilated Causal Convolutional Neural Network.

The FCN is first.

Fully connected network

In this first practice session, we'll use TensorFlow libraries, which we can quickly install from the terminal (or similarly from the anaconda navigator):

```
pip install -U tensorflow
```

We'll execute the commands from the Python (or IPython) terminal, but equally, we could execute them from a Jupyter notebook (or a different environment).

The installation could take a while – the TensorFlow library is about 200 MB in size and comes with a few dependencies.

Let's load the dataset. Here, I am assuming that you've downloaded it onto your computer:

```python
import pandas as pd
passengers = pd.read_csv(
    "passengers.csv", parse_dates=["date"]
).set_index("date")
```

Let's try naively to just use an FCN, also known as an MLP.

Let's set some imports and set a couple of global constants:

```python
import tensorflow as tf
import tensorflow.keras as keras
from tensorflow.keras.layers import Dense, Input, Dropout

DROPOUT_RATIO = 0.2
HIDDEN_NEURONS = 10

callback = tf.keras.callbacks.EarlyStopping(
    monitor='loss', patience=3
)
```

We will use these constants for our model architecture.

Dropout (or: Dilution) is a regularization technique that can help reducing overfitting. Dropout means that during training, a fraction of the connections (in our case 20%) is randomly removed.

Early stopping is another form of regularization, where the training stops as defined by certain conditions. In our case, we've stated it should stop if our loss doesn't improve three times in a row. If the model stops improving, there's no point in continuing to train it, although we may be trapped in a local minimum of the error that we might be able to escape. One of the big advantages of early stopping is that it can help us quickly see whether a model is working.

We can define our model in this function:

```
def create_model(passengers):
    input_layer = Input(len(passengers.columns))
    hiden_layer = Dropout(DROPOUT_RATIO)(input_layer)
    hiden_layer = Dense(HIDDEN_NEURONS, activation='relu')(hiden_layer)
    output_layer = Dropout(DROPOUT_RATIO)(hiden_layer)
    output_layer = Dense(1)(output_layer)
    model = keras.models.Model(
        inputs=input_layer, outputs=output_layer
    )
    model.compile(
        loss='mse',
    optimizer=keras.optimizers.Adagrad(),
        metrics=[keras.metrics.RootMeanSquaredError(), keras.metrics.
MeanAbsoluteError()])
    return model
```

With the Keras functional API, we've defined a two-layer neural network, where the hidden layer of HIDDEN_NEURONS neurons is activated by the Rectified Linear Unit (ReLU) function.

Let's split our dataset into training and test sets. We will predict the number of passengers based on passengers in the previous time period (previous month):

```
from sklearn.model_selection import train_test_split
X_train, X_test, y_train, y_test = train_test_split(
    passengers, passengers.passengers.shift(-1), shuffle=False
)
```

We'll learn based on the first 75% of the dataset – this is the default value for the test_size parameter in the train_test_split function.

We can now train our naïve FCN:

```
model = create_model(X_train)
model.fit(X_train, y_train, epochs=1000, callbacks=[callback])
```

We should get an output of the loss and the metrics at each epoch as they finish:

```
Epoch 1/1000
4/4 [==============================] - 0s 3ms/step - loss: 71752.7812 - root_m
ean_squared_error: 267.8671 - mean_absolute_error: 237.6112
Epoch 2/1000
4/4 [==============================] - 0s 2ms/step - loss: 69809.4141 - root_m
ean_squared_error: 264.2147 - mean_absolute_error: 243.9879
Epoch 3/1000
4/4 [==============================] - 0s 2ms/step - loss: 76103.9375 - root_m
ean_squared_error: 275.8694 - mean_absolute_error: 253.2610
Epoch 4/1000
4/4 [==============================] - 0s 2ms/step - loss: 76133.8672 - root_m
ean_squared_error: 275.9237 - mean_absolute_error: 253.2717
Epoch 5/1000
```

Figure 10.12: Model training

Ideally, we would see the error (loss) going down, and we'd see a low error at the end. I haven't included any code to fix the random number generator (tf.random.set_seed), so your output might differ.

We can then get the predictions for the test set like this:

```
predicted = model.predict(X_test)
```

Now, it would be good to visualize passenger predictions against the actual passenger values.

We can use this function:

```
import matplotlib.pyplot as plt

def show_result(y_test, predicted):
  plt.figure(figsize=(16, 6))
  plt.plot(y_test.index, predicted, 'o-', label="predicted")
  plt.plot(y_test.index, y_test, '.-', label="actual")
  plt.ylabel("Passengers")
  plt.legend()
```

Let's visualize our predictions then!

```
show_result(y_test, predicted)
```

Here's the graph:

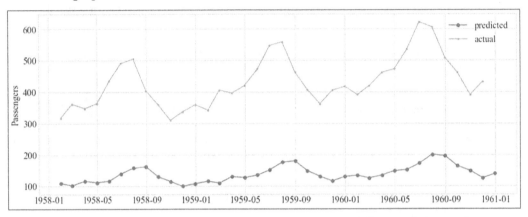

Figure 10.13: Predicted against actual airplane passengers: Naïve fully connected network

We can see that the model has learned some of the monthly variability. However, it is systematically under-predicting – it has learned the baseline from the years 1949-1958 of the training set, when far fewer passengers were traveling.

Let's make this a bit more sophisticated and better.

This first model was trained only on the immediate previous number of travelers.

As a first step, we'll include the year and the month as predictor variables. The year can be used to model the trend, while the month is coupled to the monthly variability – so this seems a natural step.

This will add month and year columns to the DataFrame based on the DateTimeIndex:

```
passengers["month"] = passengers.index.month.values
passengers["year"] = passengers.index.year.values
```

Now we can redefine our model – we need to add more input columns:

```
model = create_model(passengers)
```

And we are ready for another training round:

```
X_train, X_test, y_train, y_test = train_test_split(
    passengers, passengers.passengers.shift(-1), shuffle=False
```

```
)
model.fit(X_train, y_train, epochs=100, callbacks=[callback])
predicted = model.predict(X_test)
show_result(y_test, predicted)
```

Let's see how well the model predictions match the test data:

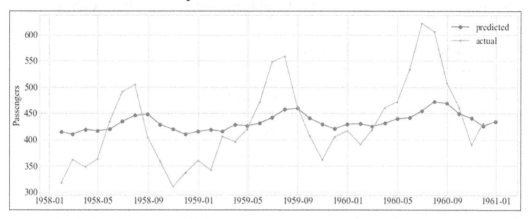

Figure 10.14: Predicted against actual airplane passengers; Fully connected network with the year and month

Please note that, because of the high number of parameters, and the randomness involved in the learning process, the outcome might differ significantly between runs. This is indeed one of the problems associated with deep learning.

This already looks much better. The year feature helped our model learn the baseline. The model has learned something about the monthly variability, but it's not enough to really approximate it.

Let's create a less naïve version. We will change a few things in this model:

- We'll add an embedding of the month feature
- We'll treat the year as a linear predictor
- We'll add the previous month's passengers to our predictions
- Finally, we'll scale our predictions based on the standard deviation in the training dataset

That was quite a mouthful. Let's go through these a bit more slowly.

We fed the months as values from 1 to 12 into our previous model. However, we could intuitively guess that January (1) and December (12) are perhaps more similar than November (11) and December. We know that there are lots of travelers in both December and January, but perhaps a much lower volume in November. We can capture these relationships based on the data.

This can be done in an embedding layer. An embedding layer is a mapping of distinct categories to real numbers. This mapping is updated as part of the network optimization.

The year is closely related to the overall trend. Each year, the number of airline passengers increases. We can model this relationship non-linearly or linearly. Here, I've decided to just model a linear relationship between the year feature and the outcome.

The relationship between the previous month's passenger numbers and those in this month is again assumed to be linear.

Finally, we can scale our predictions, similar to the inverse transformation of the standard transformation. You should remember the standard normalization from *Chapter 3, Preprocessing Time-Series*, as follows:

$$z = \frac{x - \mu}{\sigma},$$

where μ is the population mean and σ is the population standard deviation.

The inverse of this is as follows:

$$x = z\sigma + \mu$$

Our formula is as follows:

$$\hat{v}_t = p\sigma + v_{t-1},$$

where $\hat{v}_t$ are the airline passengers at time t and p is the prediction based on the embedded month and the year.

We assume the network will learn to baseline, but might not learn the scale perfectly – so we'll help out.

An illustration might help (from TensorBoard, TensorFlow's visualization toolkit):

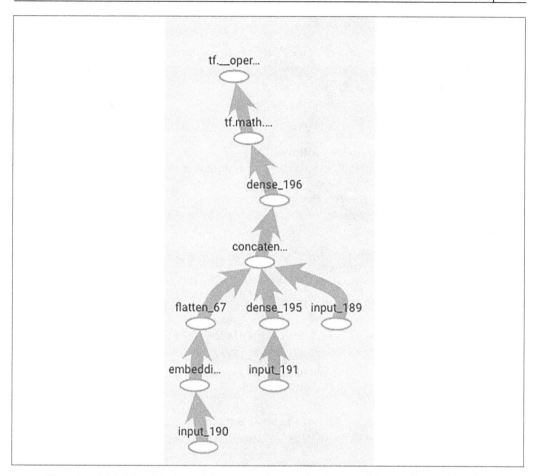

Figure 10.15: Model architecture: Fully connected network with embedding, scaling, and baseline

We can see the three inputs, one of them (the months) going through an embedding layer, and another one through a (linear) projection. They all come together (concatenate) and go through a dense layer, where another math operation is performed on top.

We'll need a few more imports first:

```
from tensorflow.keras.layers.experimental preprocessing
from tensorflow.keras.layers import Embedding, Flatten, Concatenate
from tensorflow.keras.metrics import (
  RootMeanSquaredError, MeanAbsoluteError
)
```

Now, we redefine our network as follows:

```
def create_model(train):
  scale = tf.constant(train.passengers.std())
  cont_layer = Input(shape=1)
  cat_layer = Input(shape=1)
  embedded = Embedding(12, 5)(cat_layer)
  emb_flat = Flatten()(embedded)
  year_input = Input(shape=1)
  year_layer = Dense(1)(year_input)
  hidden_output = Concatenate(-1)([emb_flat, year_layer, cont_layer])
  output_layer = keras.layers.Dense(1)(hidden_output)
  output = output_layer * scale + cont_layer
  model = keras.models.Model(inputs=[
    cont_layer, cat_layer, year_input
  ], outputs=output)
  model.compile(loss='mse', optimizer=keras.optimizers.Adam(),
    metrics=[RootMeanSquaredError(), MeanAbsoluteError()])
  return model
```

We reinitialize our model:

```
model = create_model(X_train)
```

During training and for prediction purposes, we need to feed the three types of input separately like this:

```
model.fit(
  (X_train["passengers"], X_train["year"], X_train["month"]),
  y_train, epochs=1000,
  callbacks=[callback]
)
predicted = model.predict((X_test["passengers"], X_test["year"], X_
test["month"]))
```

You might notice that the training carries on for much longer in this configuration.

This chart illustrates the fit we are achieving with our new network:

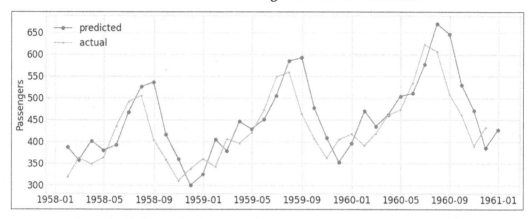

Figure 10.16: Predicted against actual airplane passenger numbers: Fully connected
network with embedding, scaling, and baseline

This again looks much better than the previous network. We leave it as an exercise to the reader to try to improve this network further.

We'll set up an RNN next.

Recurrent neural network

We discussed in the theory section that recurrent neural networks can be very good at modeling long-term relationships between points in a time-series. Let's set up an RNN.

We'll use the same dataset as before – the univariate values of airline passengers. In this case, our network is going to need a sequence of points for each training sample. At each training step, the RNN is going to be trained on points (passengers) leading up to the next passenger number.

Please note that we can use TensorFlow (or even statsmodels' `lagmat()` utility functions for this purpose (and we will use them in *Chapter 12, Case Studies*), but in this instance, we'll write this quickly ourselves.

We'll need to resample our passenger numbers thus:

```
def wrap_data(df, lookback: int):
    dataset = []
    for index in range(lookback, len(df)+1):
```

```
        features = {
            f"col_{i}": float(val) for i, val in enumerate(
                df.iloc[index-lookback:index].values
            )
        }
        row = pd.DataFrame.from_dict([features])
        row.index = [df.index[index-1]]
        dataset.append(row)
    return pd.concat(dataset, axis=0)
```

This function does the job. It goes over all the points in the dataset and takes a sequence leading up to it. The number of points in the new sequence is defined by the parameter `lookback`.

Let's put it to use:

```
LOOKBACK = 10
dataset = wrap_data(passengers, lookback=LOOKBACK)
dataset = dataset.join(passengers.shift(-1))
```

We are using a lookback of 10. I've deliberately chosen a value that is not optimal. I'm leaving it as an exercise to the reader to choose one that's better and try it out.

The last line in the code above joins the targets (lookahead 1) together with the sequences.

We are ready to define our network, but let's get the imports out of the way:

```
import tensorflow.keras as keras
from tensorflow.keras.layers import Input, Bidirectional, LSTM, Dense
import tensorflow as tf
```

The network is defined by this function:

```
def create_model(passengers):
    input_layer = Input(shape=(LOOKBACK, 1))
    recurrent = Bidirectional(LSTM(20, activation="tanh"))(input_layer)
    output_layer = Dense(1)(recurrent)
    model = keras.models.Model(inputs=input_layer, outputs=output_layer)
    model.compile(loss='mse', optimizer=keras.optimizers.Adagrad(),
        metrics=[keras.metrics.RootMeanSquaredError(), keras.metrics.
MeanAbsoluteError()])
    return model
```

It's a bidirectional LSTM network. The result of the last layer is projected linearly as our output. I've set the activation function of the LSTM to `tanh` in case you want to run this on a GPU runtime, so it will benefit from NVIDIA's GPU-accelerated library cuDNN. We are extracting the same metrics as in the previous practice.

A few more preliminaries that you should be familiar with from the previous section are as follows:

```
from sklearn.model_selection import train_test_split

callback = tf.keras.callbacks.EarlyStopping(monitor='loss', patience=3)
model = create_model(passengers)
X_train, X_test, y_train, y_test = train_test_split(
    dataset.drop(columns="passengers"),
    dataset["passengers"],
    shuffle=False
)
```

Let's train it:

```
model.fit(X_train, y_train, epochs=1000, callbacks=[callback])
```

The result looks quite good already – even though we made a few sub-optimal choices:

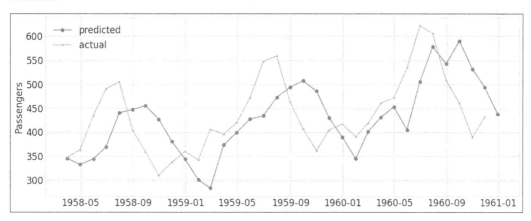

Figure 10.17: Recurrent neural network for passenger forecasts

Given how easy this was to set up, this already looks very promising.

Let's now try a causal ConvNet!

Dilated causal convolutional neural network

This example is based on Krist Papadopoulos's SeriesNet implementation of the paper *"Conditional Time-Series Forecasting with Convolutional Neural Networks,"* by Anastasia Borovykh and others.

We'll implement this model together and we'll apply it to two datasets to see how it does out of the box. I won't be going through tweaking the data and architecture in this example.

First, the imports:

```
import numpy as np
import pandas as pd
from keras.layers import Conv1D, Input, Add, Activation, Dropout
from keras.models import Sequential, Model
from keras.layers.advanced_activations import LeakyReLU, ELU
from keras import optimizers
import tensorflow as tf
```

What is perhaps surprising is how easy it is to do a causal convolution in TensorFlow. Conv1D comes with a parameter, padding, which can be specified as 'causal'. This simply pads the layer's input with zeros according to the causal nature, where output at time t only depends on the previous time steps, <t. Please refer to the discussion in the ConvNets section in this chapter.

This means that we can predict the values of early time steps in the frame.

The main idea of this network is a residual block with causal convolutions. This code segment constructs the corresponding network architecture:

```
def DC_CNN_Block(nb_filter, filter_length, dilation):
    def f(input_):
        residual =    input_
        layer_out =    Conv1D(
            filters=nb_filter, kernel_size=filter_length,
            dilation_rate=dilation,
            activation='linear', padding='causal', use_bias=False
        )(input_)
        layer_out =    Activation('selu')(layer_out)
        skip_out =    Conv1D(1, 1, activation='linear', use_bias=False)
(layer_out)
        network_in =    Conv1D(1, 1, activation='linear', use_bias=False)
(layer_out)
```

```
        network_out = Add()([residual, network_in])
        return network_out, skip_out
    return f
```

I've simplified this a bit to make it easier to read.

The network itself just stacks these layers as a SkipNet follows up with a convolution:

```
def DC_CNN_Model(length):
    input = Input(shape=(length,1))
    l1a, l1b = DC_CNN_Block(32, 2, 1)(input)
    l2a, l2b = DC_CNN_Block(32, 2, 2)(l1a)
    l3a, l3b = DC_CNN_Block(32, 2, 4)(l2a)
    l4a, l4b = DC_CNN_Block(32, 2, 8)(l3a)
    l5a, l5b = DC_CNN_Block(32, 2, 16)(l4a)
    l6a, l6b = DC_CNN_Block(32, 2, 32)(l5a)
    l6b = Dropout(0.8)(l6b)
    l7a, l7b = DC_CNN_Block(32, 2, 64)(l6a)
    l7b = Dropout(0.8)(l7b)
    l8 =   Add()([l1b, l2b, l3b, l4b, l5b, l6b, l7b])
    l9 =   Activation('relu')(l8)
    l21 =  Conv1D(1, 1, activation='linear', use_bias=False)(l9)
    model = Model(inputs=input, outputs=l21)
    model.compile(loss='mae', optimizer=optimizers.Adam(),
 metrics=['mse'])
    return model
```

This is for a univariate time-series. For a multivariate time-series, some changes are necessary, which we won't cover here.

Let's forecast passenger numbers again. We'll load the DataFrame as in the previous practice sections:

```
passengers = pd.read_csv(
  "passengers.csv", parse_dates=["date "]
).set_index("date")
```

We'll split this again into test and training sets:

```
from sklearn.model_selection import train_test_split
X_train, X_test, y_train, y_test = train_test_split(
    passengers.passengers, passengers.passengers.shift(-1),
shuffle=False
)
```

We'll train the model with this function:

```
def fit_model(timeseries):
    length = len(timeseries)-1
    model = DC_CNN_Model(length)
    model.summary()
    X = timeseries[:-1].reshape(1,length, 1)
    y = timeseries[1:].reshape(1,length, 1)
    model.fit(X, y, epochs=3000, callbacks=[callback])
    return model
```

This function will do the forecast for us:

```
def forecast(model, timeseries, horizon: int):
    length = len(timeseries)-1
    pred_array = np.zeros(horizon).reshape(1, horizon, 1)
    X_test_initial = timeseries[1:].reshape(1,length,1)
    pred_array[: ,0, :] = model.predict(X_test_initial)[:, -1:, :]
    for i in range(horizon-1):
        pred_array[:, i+1:, :] = model.predict(
            np.append(
                X_test_initial[:, i+1:, :],
                pred_array[:, :i+1, :]
            ).reshape(1, length, 1))[:, -1:, :]
    return pred_array.flatten()
```

The forecast is created by predicting the immediate next future value based on the previous predictions. The parameter `horizon` is the forecast horizon.

We'll put this together as a single function for convenience:

```
def evaluate_timeseries(series, horizon: int):
    model = fit_model(series)
    pred_array = forecast(model, series, horizon)
    return pred_array, model
```

Now we are ready for training. We'll run everything like this:

```
HORIZON = len(y_test)
predictions, model = evaluate_timeseries(
    X_train.values.reshape(-1, 1), horizon= HORIZON
)
```

The model is very deep, but is not that big in terms of parameters because of the convolutions. We'll see that there are 865 trainable parameters.

The model fit is not that good though, neither in MSE nor does that graph look very impressive either:

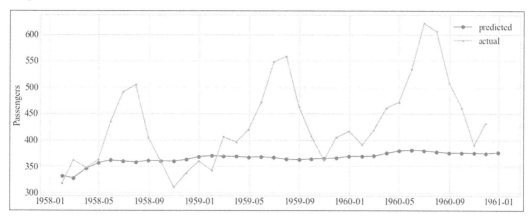

Figure 10.18: ConvNet prediction of passenger numbers

This graph can be produced by running show_result(y_test[:HORIZON], predictions[:HORIZON], "Passengers").

This highlights the fact that each model has its strengths and weaknesses and that without adapting a model to our dataset and careful preprocessing, we won't be able to get a good performance. It's left as an exercise to the reader to try and tweak this model.

Summary

In this chapter, I've introduced many deep learning concepts relevant to time-series, and we've discussed many architectures and algorithms, such as autoencoders, InceptionTime, DeepAR, N-BEATS, ConvNets, and a few transformer architectures. Deep learning algorithms are indeed coming very close to the state of the art in time-series, and it's an exciting area of research and application.

In the practice section, I implemented a fully connected feedforward network and then an RNN before taking a causal ConvNet for a ride.

In *Chapter 12*, *Multivariate Forecasting*, we'll do some more deep learning, including a Transformer model and an LSTM.

11

Reinforcement Learning for Time-Series

Reinforcement learning is a widely successful paradigm for control problems and function optimization that doesn't require labeled data. It's a powerful framework for experience-driven autonomous learning, where an agent interacts directly with the environment by taking actions and improves its efficiency by trial and error. Reinforcement learning has been especially popular since the breakthrough of the London-based Google-owned company DeepMind in complex games.

In this chapter, we'll discuss a classification of **reinforcement learning (RL)** approaches in time-series specifically economics, and we'll deal with the accuracy and applicability of RL-based time-series models.

We'll start with core concepts and algorithms in RL relevant to time-series and we'll talk about open issues and challenges in current deep RL models.

I am going to cover the following topics:

- Introduction to Reinforcement Learning
- Reinforcement Learning for Time-Series
- Bandit algorithms
- Deep Q-Learning
- Python Practice

Let's start with an introduction to reinforcement learning and the main concepts.

Introduction to reinforcement learning

Reinforcement learning is one of the main paradigms in machine learning alongside supervised and unsupervised methods. A major distinction is that supervised or unsupervised methods are passive, responding to changes, whereas RL is actively changing the environment and seeking out new data. In fact, from a machine learning perspective, reinforcement learning algorithms can be viewed as alternating between finding good data and doing supervised learning on that data.

Computer programs based on reinforcement learning have been breaking through barriers. In a watershed moment for artificial intelligence, in March 2016, DeepMind's AlphaGo defeated the professional Go board game player Lee Sedol. Previously, the game of Go was considered to be a hallmark of human creativity and intelligence, too complex to be learned by a machine.

It has been argued that it is edging us closer toward **Artificial General Intelligence** (**AGI**). For example, in their paper *"Reward is enough"* (2021), David Silver, Satinder Singh, Doina Precup, and Richard S. Sutton argue that reward-orientated learning is enough to acquire knowledge, learn, perceive, socialize, understand and produce language, generalize, and imitate. More emphatically, they state that reinforcement learning agents could constitute a solution to AGI.

Artificial General Intelligence (**AGI**) is the hypothetical ability of an intelligent agent to understand or learn any intellectual task that would require intelligence. What is **intelligence** though? Often this is defined as anything humans can do or would consider hard. According to Turing Award winning computer scientist John McCarthy (*"What Is AI?"* 1998), *"intelligence is the computational part of the ability to achieve goals in the world."*

In reinforcement learning, an agent interacts with the environment through actions and gets feedback in the shape of rewards. Contrary to the situation in supervised learning, no labeled data is available, but rather the environment is explored and exploited on the basis of the expectation of cumulative rewards. This feedback cycle of action and reward is illustrated in this diagram:

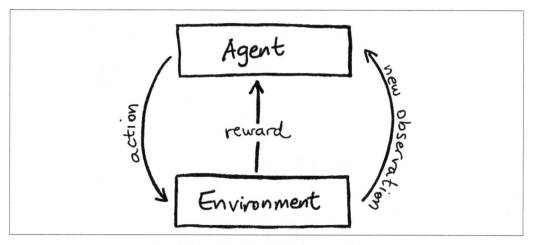

Figure 11.1: Feedback loop in reinforcement learning

Reinforcement learning is concerned with the objective of reward maximization. By interacting with the environment, the agent gets feedback and learns to take better actions. By optimizing the cumulative reward, the agent develops goal-directed behavior.

 Reinforcement learning (RL) is an approach where an agent interacts directly with the environment by taking actions. The agent learns through trial and error to maximize the reward.

If you've read *Chapter 8, Online Learning for Time-Series*, you might be confused about the difference between reinforcement learning and online learning, and it might be worthwhile to consider the two approaches in comparison. Some of the most prominent algorithms for reinforcement learning, Q-learning and Temporal Difference (TD) learning, to just name a couple of examples, are online algorithms, the way they update the value function.

However, reinforcement learning doesn't focus on predictions, but on the interaction with the environment. In **online learning**, information is processed continuously, and the problem is clearly defined in terms of what's correct and what's incorrect. In reinforcement learning, the goal is the optimization of a delayed reward over a number of steps interacting with the environment. This is the main difference between the two approaches, although there are many particular details proponents of each technique would claim as theirs. Some of these we'll discuss later in this chapter, such as exploration versus exploitation and experience replay.

A reinforcement problem is defined by three main components: the environment ε, the agent A, and the cumulative objective. The agent is a decision-making entity that can observe the current state of the environment and takes an action. By performing an action $a \in A$, the agent transitions from state to state, $s_t \rightarrow s_{t+1}$. Executing an action in a specific state provides the agent with a reward, which is a numerical score. The reward is an instantaneous measurement of progress towards a goal.

The environment is in a certain state that depends on some combination of the current state and the action taken, although some of the changes could be random. It's the agent's objective to maximize a cumulative reward function. This cumulative reward objective can be the sum of rewards over a number of steps, a discounted sum, or the average reward over time.

More formally, an agent in the context of RL is a system (or program) that receives an observation O_t of the environment at time t and outputs an action $A_t = \alpha(H_t)$ given its history of experiences $H_t = (O_1, A_1), \ldots, (O_{t-1}, A_{t-1}), (O_t, A_t)$.

Meanwhile, an environment is another system. It receives an action A_t at time t and changes its state in accordance with the history of actions and past states and a random process η. The state is accessible to the agent to a certain degree, and to simplify we can state: $O_{t+1} = \epsilon(H_t, A_t, \eta_t)$.

Finally, a reward is a scalar observation that is emitted at every time step t by the environment that provides momentaneous feedback to the agent on how well it is doing.

At the core of the reinforcement learning agent is a model that estimates the value of an environmental state or suggests the next action in the environment. These are the two main categories of reinforcement learning: in **value-based** learning, a model approximates the outcomes of actions or the value of environmental states with a value function (a model) and the action selection reduces to take the action with the best expected outcome. In **policy-based** learning, we focus on the more direct goal of choosing the action by predicting an action from the environmental state.

There's another twist to reinforcement learning: the **exploration versus exploitation dilemma**. You can decide to keep doing what you know works best (exploitation) or try out new avenues (exploration). Trying out new things will probably lead to worse results in the short run but might teach you important lessons that you can draw from in the future.

A simple approach to balance the two against each other is **epsilon-greedy**. This is a simple method to balance exploration and exploitation by choosing between exploration and exploitation randomly: either we follow our model's advice, or we don't. Epsilon is the parameter for the probability that we do an action that's not recognized as the best by the model; the higher epsilon, the more random the model's actions.

Deep reinforcement learning (**DRL**) techniques are a subset of reinforcement learning methods, where the model is a deep neural network (or, in a looser sense, a multilayer-perceptron).

In the next section, we'll look into how RL can be applied to time-series!

Reinforcement Learning for Time-Series

Reinforcement Learning (RL) can and has been applied to time-series, however, the problem has to be framed in a certain way. For reinforcement learning, we need to have significant feedback between predictions and ongoing (actions) of the system.

In order to apply RL to time-series forecasting or predictions, the prediction has to condition an action, therefore the state evolution depends on the current state and the agent's action (and randomness). Hypothetically, rewards could be a performance metric about the accuracy of predictions. However, the consequences of good or bad predictions do not affect the original environment. Essentially this corresponds to a supervised learning problem.

More meaningfully, if we want to frame our situation as an RL problem, the state of the systems should be affected by the agents' decisions. For instance, in the case of interacting with the stock market, we would buy or sell based on predictions of the movements and include something that we influence such as our portfolio and funds in the state, or (only really if we are a market maker) the influence we have over the stock movements.

In summary, RL is very apt for dealing with processes that change over time, although RL deals with those that can be controlled or influenced. A core application for time-series is in industrial processes and control – this was in fact already pointed out by Box and Jenkins in their classic book "*Time-Series Analysis: Forecasting and Control*").

There are lots of applications that we could think of for reinforcement learning. Trading on the stock market is a major driver of business growth, and the presence of uncertainty and risk recommends it as a reinforcement learning use case. In pricing, for example in insurance or retail, reinforcement learning can help explore the space of value proposition for customers that would yield high sales, while optimizing the margin. Finally auction mechanisms, for example online bidding for advertisements, are another domain. In auctions, reinforcement agents have to develop responses in the presence of other players.

Let's go more into detail about a few algorithms – first, bandits.

Bandit algorithms

A **Multi-Armed Bandit** (**MAB**) is a classic reinforcement learning problem, in which a player is faced with a slot machine (bandit) that has k levers (arms), each with a different reward distribution. The agent's goal is to maximize its cumulative reward on a trial-by-trial basis. Since MABs are a simple but powerful framework for algorithms that make decisions over time under uncertainty, a large number of research articles have been dedicated to them.

Bandit learning refers to algorithms that aim to optimize a single unknown stationary objective function. An agent chooses an action from a set of actions $a \in A$. The environment reveals reward $r_t(a)$ of the chosen action at time t. As information is accumulated over multiple rounds, the agent can build a good representation of the value (or reward) distribution for each arm, $Q(a)$.

Therefore, a good policy might converge so that the choice of arm becomes optimal. According to one policy, **UCB1** (published by Peter Auer, Nicolò Cesa-Bianchi, and Paul Fischer, *"Finite-Time Analysis of the Multi-Armed Bandit Problem"*, 2002), given the expected values for each action, the action is chosen that maximizes this criterion:

$$argmax_{a \in A} \ Q(a) + \sqrt{\frac{2 \ln t}{N_t(a)}}$$

The second term refers to the upper confidence bound of the reward values based on the information we have accumulated. Here, t refers to the number of iterations so far, the time step, and $N_t(a)$ to the number of times action a has been executed so far. This means that the nominator in the equation increases logarithmically with time and the denominator increases each time we receive reward information from the action.

When the available rewards are binary (win or lose, yes or no, charge or no charge) then this can be described by a Beta distribution. The Beta distribution takes two parameters, α and β, for wins and losses, respectively. The mean value is $\frac{\alpha}{\alpha + \beta}$.

In **Thompson sampling**, we sample from the Beta distribution of each action (arm) and choose the action with the highest estimated return. The Beta distribution narrows with the number of tries, therefore actions that have been tried infrequently have wide distributions. Therefore, Beta sampling models the estimated mean reward and the level of confidence in the estimate. In **Dirichlet sampling**, instead of sampling from a Beta distribution, we are sampling from a Dirichlet distribution (also called multivariate Beta distribution).

Contextual bandits incorporate information about the environment for updating the reward expectation. If you think about ads, this contextual information could be if the ad is about traveling. The advantage of contextual bandits is agents can encode much richer information about the environment.

In contextual bandits, an agent chooses an arm, the reward $r_t(a)$ is revealed, and the agent's expectation of the reward is updated, but with context features: $Q(a, x)$, where x is a set of features encoding the environment. In many implementations, the context is often restricted to discrete values, however, at least in theory, they could be either categorical or numerical. The value function could be any machine learning algorithm such as a neural network (NeuralBandit) or a random forest (BanditForest).

Bandits find applications, among other fields, in information retrieval models such as recommender and ranking systems, which are employed in search engines or on consumer websites. The **probability ranking principle** (PRP; from S.E. Robertson's article *"The probability ranking principle in IR"*, 1977) forms the theoretical basis for probabilistic models, which have been dominating IR. The PRP states that articles should be ranked in decreasing order of relevance probability. This is what we'll go through in an exercise in the practice section.

Let's delve into Q-learning and deep Q-learning now.

Deep Q-Learning

Q-learning, introduced by Chris Watkins in 1989, is an algorithm to learn the value of an action in a particular state. Q-learning revolves around representing the expected rewards for an action taken in a given state.

The expected reward of the state-action combination SxA is approximated by the Q function:

$$Q: SxA \rightarrow \mathbb{R}$$

Q is initialized to a fixed value, usually at random. At each time step t, the agent selects an action $a_t \in A$ and sees a new state of the environment s_{t+1} as a consequence and receives a reward.

The value function Q can then be updated according to the Bellman equation as the weighted average of the old value and the new information:

$$Q^{new}(s_t, a_t) \leftarrow (1 - \alpha)Q(s_t, a_t) + \alpha(r_t + \gamma \max_a \{ Q(s_{t+1}, a) \})$$

The weight is by α, the learning rate – the higher the learning rate, the more adaptive the Q-function. The discount factor $0 < \gamma \leq 1$ is weighting the rewards by their immediacy – the higher γ, the more impatient (myopic) the agent becomes.

$(1 - \alpha)Q(s_t, a_t)$ represents the current reward. αr_t is the reward obtained by s_t weighted by learning rate α, and $\alpha \gamma \max\{ Q(s_{t+1}, a) \}$ is the weighted maximum reward that can be obtained from state s_{t+1}.

This last part can be recursively broken down into simpler sub-problems like this:

$$\gamma \max\{ Q(s_{t+1}, a) \}) = \max_a \gamma Q(s_{t+1}, a) + \max_{a'} \gamma^2 Q(s_{t+2}, a')$$

In the simplest case, Q can be a lookup table, called a Q-table.

In 2014, Google DeepMind patented an algorithm called **deep Q-learning**. This algorithm was introduced in the Nature paper "*Human-level control through deep reinforcement learning*" with an application in Atari 2600 games.

In Deep Q-learning, a neural network is used for the Q-function as a nonlinear function approximator. They used a convolutional neural network to learn expected rewards from pixel values. They introduced a technique called **experience replay** to update Q over a randomly drawn sample of prior actions. This is done to reduce the learning instability of the Q updates.

Q-learning can be shown in pseudocode roughly like this:

```python
import numpy as np
memory = []
for episode in range(N):
  for ts in range(T):
    if eps np.random.random() > epsilon:
      a = A[np.argmax([Q(a) for a in A])]
    else:
      a = np.random.choice(A)
    r, s_next = env.execute(a)
    memory.append((s, a, r, s_next))
    learn(np.random.choice(memory, L)
```

This implements an epsilon-greedy policy by which a random (exploratory) choice is made according to the probability epsilon. A few more variables are assumed given. The handle for the environment, env, allows us to execute an action. We have a learning function, which applies gradient descent on the Q-function to learn better values according to the Bellman equation. The parameter L is the number of previous values that are used for learning.

The memory replay part is obviously simplified. In actuality, we would have a maximum size of the memory, and, once the memory capacity is reached, we would replace old associations of states, actions, and rewards with new ones.

We'll put some of this into practice now.

Python Practice

Let's get into modeling. We'll start by giving some recommendations for users using MABs.

Recommendations

For this example, we'll take joke preferences by users, and we'll use them to simulate feedback on recommended jokes on our website. We'll use this feedback to tune our recommendations. We want to select the 10 best jokes to present to people visiting our site. The recommendations are going to be produced by 10 MABs that each have as many arms as there are jokes.

This is adapted from an example from the `mab-ranking` library on GitHub by Kenza-AI.

It's a handy library that comes with implementations of different bandits. I've simplified the installation of this library in my fork of the library, so we'll be using my fork here:

```
pip install git+https://github.com/benman1/mab-ranking
```

After this is finished, we can get right to it!

We'll download the `jester` dataset with joke preferences from S3. Here's the location:

```
URL = 'https://raw.githubusercontent.com/PacktPublishing/Machine-
Learning-for-Time-Series-with-Python/main/chapter11/jesterfinal151cols.
csv'
```

We'll download them using pandas:

```
import pandas as pd
jester_data = pd.read_csv(URL, header=None)
```

We'll make some cosmetic adjustments. The rows refer to users, the columns to jokes. We can make this clearer:

```
jester_data.index.name = "users"
```

The encoding of choices is a bit weird, so we'll fix this as well:

```
for col in jester_data.columns:
    jester_data[col] = jester_data[col].apply(lambda x: 0.0 if x>=99 or
x<7.0 else 1.0)
```

So either people chose the joke or they didn't. We'll get rid of people who didn't choose any joke at all:

```
jester_data = jester_data[jester_data.sum(axis=1) > 0]
```

Our dataset looks like this now:

	0	1	2	3	4	5	6	7	8	9	...	141	142	143	144	145	146	147	148	149	150
users																					
0	1.0	0.0	0.0	0.0	0.0	0.0	0.0	0.0	0.0	0.0	...	0.0	0.0	0.0	0.0	0.0	0.0	0.0	0.0	0.0	0.0
1	1.0	0.0	0.0	0.0	0.0	0.0	0.0	1.0	1.0	0.0	...	0.0	0.0	0.0	0.0	0.0	0.0	0.0	0.0	0.0	0.0
2	1.0	0.0	0.0	0.0	0.0	0.0	0.0	0.0	0.0	0.0	...	0.0	0.0	0.0	0.0	0.0	0.0	0.0	0.0	0.0	0.0
3	1.0	0.0	0.0	0.0	0.0	0.0	0.0	0.0	0.0	0.0	...	0.0	0.0	0.0	0.0	0.0	0.0	0.0	0.0	0.0	0.0
4	1.0	0.0	0.0	0.0	0.0	0.0	0.0	0.0	0.0	0.0	...	0.0	0.0	0.0	0.0	0.0	0.0	0.0	0.0	0.0	0.0
...	...	...	...	...	...	...	...	...	...	...	...	...	...	...	...	...	...	...	...	...	...
50687	1.0	0.0	0.0	0.0	0.0	0.0	0.0	0.0	0.0	0.0	...	0.0	0.0	0.0	0.0	0.0	0.0	0.0	0.0	0.0	0.0
50688	1.0	0.0	0.0	0.0	0.0	0.0	0.0	0.0	0.0	0.0	...	0.0	0.0	0.0	0.0	0.0	0.0	0.0	0.0	0.0	0.0
50689	1.0	0.0	0.0	0.0	0.0	0.0	0.0	0.0	0.0	0.0	...	0.0	0.0	0.0	0.0	0.0	0.0	0.0	0.0	0.0	0.0
50690	1.0	0.0	0.0	0.0	0.0	0.0	0.0	0.0	0.0	0.0	...	0.0	0.0	0.0	0.0	0.0	0.0	0.0	0.0	0.0	0.0
50691	1.0	0.0	0.0	0.0	0.0	0.0	0.0	0.0	0.0	0.0	...	1.0	1.0	1.0	1.0	1.0	1.0	1.0	1.0	1.0	1.0

49961 rows × 151 columns

Figure 11.2: Jester dataset

We'll set up our bandits as follows:

```
from mab_ranking.bandits.rank_bandits import IndependentBandits
from mab_ranking.bandits.bandits import DirichletThompsonSampling

independent_bandits = IndependentBandits(
    num_arms=jester_data.shape[1],
    num_ranks=10,
    bandit_class=DirichletThompsonSampling
)
```

We choose independent bandits with Thompson sampling from the Beta distribution. We recommend the best 10 jokes.

We can then start our simulation. Our hypothetical website has lots of visits, and we'll get feedback on the 10 jokes that we'll display as chosen by our independent bandits:

```
from tqdm import trange
num_steps = 7000
hit_rates = []
for _ in trange(1, num_steps + 1):
```

```
    selected_items = set(independent_bandits.choose())
    # Pick a users choices at random
    random_user = jester_data.sample().iloc[0, :]
    ground_truth = set(random_user[random_user == 1].index)
    hit_rate = len(ground_truth.intersection(selected_items)) /
len(ground_truth)
    feedback_list = [1.0 if item in ground_truth else 0.0 for item in
selected_items]
    independent_bandits.update(selected_items, feedback_list)
    hit_rates.append(hit_rate)
```

We are simulating 7,000 iterations (visits). At each visit, we'll change our choices according to the updated reward expectations.

We can plot the hit rate, the jokes that users are selecting, as follows:

```
import matplotlib.pyplot as plt
stats = pd.Series(hit_rates)
plt.figure(figsize=(12, 6))
plt.plot(stats.index, stats.rolling(200).mean(), "--")
plt.xlabel('Iteration')
plt.ylabel('Hit rate')
```

I've introduced a rolling average (over 200 iterations) to get a smoother graph:

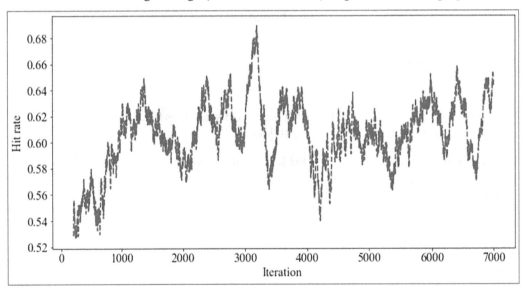

Figure 11.3: Hit rate over time (Dirichlet sampling)

The mab-ranking library supports contextual information, so we can try out giving additional information. Let's imagine this information as different user groups (cohorts). We could think of users who use different search or filter functionality on our imaginary website, say "newest jokes" or "most popular." Alternatively, they could be from different regions. Or it could be a timestamp category that corresponds to the time of the day of visits of users to our website.

Let's supply the categorical user group information, the context. We'll cluster users by their preferences, and we'll use the clusters as context:

```
from sklearn.cluster import KMeans
from sklearn.preprocessing import StandardScaler
scaler = StandardScaler().fit(jester_data)
kmeans = KMeans(n_clusters=5, random_state=0).fit(scaler.
transform(jester_data))
contexts = pd.Series(kmeans.labels_, index=jester_data.index)
```

This creates 5 user groups.

We'll have to reset our bandits:

```
independent_bandits = IndependentBandits(
    num_arms=jester_data.shape[1],
    num_ranks=10,
    bandit_class=DirichletThompsonSampling
)
```

Then, we can redo our simulation. Only now, we'll supply the user context:

```
hit_rates = []
for _ in trange(1, num_steps + 1):
    # Pick a users choices at random
    random_user = jester_data.sample().iloc[0, :]
    context = {"previous_action": contexts.loc[random_user.name]}
    selected_items = set(independent_bandits.choose(
        context=context
    ))
    ground_truth = set(random_user[random_user == 1].index)
    hit_rate = len(ground_truth.intersection(selected_items)) /
len(ground_truth)
    feedback_list = [1.0 if item in ground_truth else 0.0 for item in
selected_items]
```

```
    independent_bandits.update(selected_items, feedback_list,
context=context)
    hit_rates.append(hit_rate)
```

We can visualize the hit rate over time again:

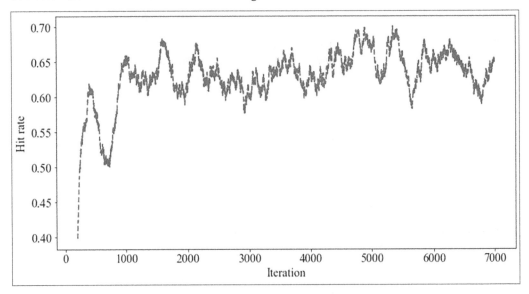

Figure 11.4: Hit rate over time (Dirichlet sampling with context)

We can see that the hit rate is a bit higher than before.

This model ignores the order of the recommended jokes on our hypothetical website. There are other bandit implementations that model the ranks.

I'll leave it to the reader to play around with this more. A fun exercise is to create a probabilistic model of reward expectations.

In the next section, we'll be playing around with a deep Q-learning trading bot. This is a more intricate model and will require more attention. We'll apply this to cryptocurrency trading.

Trading with DQN

This is based on a tutorial of the TensorTrade library, which we'll use in this example. TensorTrade is a framework for building, training, evaluating, and deploying robust trading algorithms using reinforcement learning.

TensorTrade relies on existing tools such as OpenAI Gym, Keras, and TensorFlow to enable fast experimentation with algorithmic trading strategies. We'll install it with pip as usual. We'll make sure we install the latest version from GitHub:

```
pip install git+https://github.com/tensortrade-org/tensortrade.git
```

We could also install the `ta` library, which can provide additional signals useful for trading, but we'll leave this out here.

Let's get a few imports out of the way:

```
import pandas as pd
import tensortrade.env.default as default
from tensortrade.data.cdd import CryptoDataDownload
from tensortrade.feed.core import Stream, DataFeed
from tensortrade.oms.exchanges import Exchange
from tensortrade.oms.services.execution.simulated import execute_order
from tensortrade.oms.instruments import USD, BTC, ETH
from tensortrade.oms.wallets import Wallet, Portfolio
from tensortrade.agents import DQNAgent
%matplotlib inline
```

These imports concern utilities for the (simulated) exchange, the portfolio, and the environment. Further, there are utilities for data loading and feeding it into the simulation, constants for currency conversion, and finally, there's a deep Q-agent, which consists of a Deep Q-Network (DQN).

Please note that the matplotlib magic command (`%matplotlib inline`) is needed for the Plotly charts to show up as expected.

As a first step, let's load a dataset of historical cryptocurrency prices:

```
cdd = CryptoDataDownload()

data = cdd.fetch("Bitstamp", "USD", "BTC", "1h")
data.head()
```

This dataset consists of hourly Bitcoin prices in US dollars. It looks like this:

	date	unix	open	high	low	close	volume
0	2018-05-15 06:00:00	1526364000	8733.86	8796.68	8707.28	8740.99	559.93
1	2018-05-15 07:00:00	1526367600	8740.99	8766.00	8721.11	8739.00	273.58
2	2018-05-15 08:00:00	1526371200	8739.00	8750.27	8660.53	8728.49	917.79
3	2018-05-15 09:00:00	1526374800	8728.49	8754.40	8701.35	8708.32	182.62
4	2018-05-15 10:00:00	1526378400	8708.32	8865.00	8695.11	8795.90	1260.69

Figure 11.5: Crypto dataset

We'll add a relative strength indicator signal, a technical indicator for the financial markets. It measures the strength or weakness of a market by comparing the closing prices of a recent trading period. We'll also add a **moving average convergence/ divergence (MACD)** indicator, which is designed to reveal changes in the strength, direction, momentum, and duration of a trend.

These two are defined as follows:

```
def rsi(price: Stream[float], period: float) -> Stream[float]:
    r = price.diff()
    upside = r.clamp_min(0).abs()
    downside = r.clamp_max(0).abs()
    rs = upside.ewm(alpha=1 / period).mean() / downside.ewm(alpha=1 /
period).mean()
    return 100*(1 - (1 + rs) ** -1)
```

```
def macd(price: Stream[float], fast: float, slow: float, signal: float)
-> Stream[float]:
    fm = price.ewm(span=fast, adjust=False).mean()
    sm = price.ewm(span=slow, adjust=False).mean()
    md = fm - sm
    signal = md - md.ewm(span=signal, adjust=False).mean()
    return signal
```

Alternatively, here we could be using trading signals from the ta library.

We'll now set up the feed that goes into our decision making:

```
features = []
for c in data.columns[1:]:
    s = Stream.source(list(data[c]), dtype="float").rename(data[c].
name)
    features += [s]

cp = Stream.select(features, lambda s: s.name == "close")
```

We are selecting the closing price as a feature.

Now, we'll add our indicators as additional features:

```
features = [
    cp.log().diff().rename("lr"),
    rsi(cp, period=20).rename("rsi"),
    macd(cp, fast=10, slow=50, signal=5).rename("macd")
]

feed = DataFeed(features)
feed.compile()
```

Aside from RSI and MACD, we are also adding a trend indicator (LR).

We can have a look at the first five lines from the data feed:

```
for i in range(5):
    print(feed.next())
```

Here is what our trading signal features look like:

```
{'lr': nan, 'rsi': nan, 'macd': 0.0}
{'lr': -0.00022768891842694927, 'rsi': 0.0, 'macd': -0.1891859774210995}
{'lr': -0.0012033785355889393, 'rsi': 0.0, 'macd': -1.2726616061000744}
{'lr': -0.0023134975946028646, 'rsi': 0.0, 'macd': -3.6577343503541435}
{'lr': 0.01000681330867259, 'rsi': 74.26253567956897, 'macd': 3.7087743627464844}
```

Figure 11.6: Data feed for trading

Let's set up the broker:

```
bitstamp = Exchange("bitstamp", service=execute_order)(
    Stream.source(list(data["close"]), dtype="float").rename("USD-BTC")
)
```

The exchange is the interface that will let us execute orders. An exchange needs a name, an execution service, and streams of price data. Currently, TensorTrade supports a simulated execution service using simulated or stochastic data.

Now we need a portfolio:

```
portfolio = Portfolio(USD, [
    Wallet(bitstamp, 10000 * USD),
    Wallet(bitstamp, 10 * BTC)
])
```

A portfolio can be any combination of exchanges and instruments that the exchange supports.

TensorTrade includes lots of monitoring tools, called renderers, which can be attached to the environment. They can draw a chart (`PlotlyTradingChart`) or log to a file (`FileLogger`), for example. Here's our setup:

```
renderer_feed = DataFeed([
    Stream.source(list(data["date"])).rename("date"),
    Stream.source(list(data["open"]), dtype="float").rename("open"),
    Stream.source(list(data["high"]), dtype="float").rename("high"),
    Stream.source(list(data["low"]), dtype="float").rename("low"),
    Stream.source(list(data["close"]), dtype="float").rename("close"),
    Stream.source(list(data["volume"]), dtype="float").rename("volume")
])
```

Finally, here's the trading environment, which is an instance of the OpenAI Gym (the OpenAI Gym provides a wide variety of simulated environments):

```
env = default.create(
    portfolio=portfolio,
    action_scheme="managed-risk",
    reward_scheme="risk-adjusted",
    feed=feed,
    renderer_feed=renderer_feed,
    renderer=default.renderers.PlotlyTradingChart(),
    window_size=20
)
```

You might be familiar with Gym environments if you've done reinforcement learning before.

Let's check the Gym feed:

```
env.observer.feed.next()
```

Here's what comes through:

```
{'external': {'lr': nan, 'macd': 0.0, 'rsi': nan},
 'internal': {'bitstamp:/BTC:/free': 10.0,
  'bitstamp:/BTC:/locked': 0.0,
  'bitstamp:/BTC:/total': 10.0,
  'bitstamp:/BTC:/worth': 87409.9,
  'bitstamp:/USD-BTC': 8740.99,
  'bitstamp:/USD:/free': 10000.0,
  'bitstamp:/USD:/locked': 0.0,
  'bitstamp:/USD:/total': 10000.0,
  'net_worth': 97409.9},
 'renderer': {'close': 8740.99,
  'date': Timestamp('2018-05-15 06:00:00'),
  'high': 8796.68,
  'low': 8707.28,
  'open': 8733.86,
  'volume': 559.93}}
```

Figure 11.7: Environment data feed for trading bot

This is what the trading bot will be able to rely on for making decisions on executing trades.

Now we can set up and train our DQN trading agent:

```
agent = DQNAgent(env)
agent.train(n_steps=200, n_episodes=2, save_path="agents/")
```

It might be a good point here to explain the difference between an epoch and an episode. Readers will probably be familiar with an epoch, which is a single pass over all training examples, whereas an episode is specific to the context of reinforcement learning. An episode is a sequence of states, actions, and rewards, which ends with a terminal state.

We get lots of plotting output from our renderer. Here's the first output I got (yours might differ a bit):

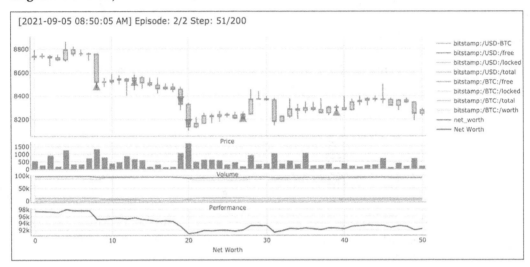

Figure 11.8: PlotlyPlotRenderer – Episode 2/2 Step 51/200

This plot gives an overview of the market operations of our trading bot. The first subplot shows the up and down movements of the prices. Then the second subplot charts volumes of stock in the portfolio, and in the bottom-most subplot, you can see the portfolio net worth.

If you want to see the net worth over time (not only the first snapshot as above), you can plot this as well:

```
performance["net_worth"].plot()
```

Here's the portfolio net worth over time:

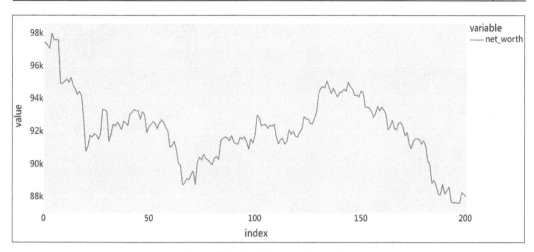

Figure 11.9: Portfolio worth over time

It looks like our trading bot could need some more training before getting let loose in the wild. I made a loss, so I am happy there wasn't real money on the line.

This is all folks. Let's summarize.

Summary

While online learning, which we talked about in *Chapter 8, Online Learning for Time-Series* is tackling traditional supervised learning, reinforcement learning tries to deal with the environment. In this chapter, I've introduced reinforcement learning concepts relevant to time-series, and we've discussed many algorithms, such as deep Q-learning and **MABs**.

Reinforcement learning algorithms are very useful in certain contexts like recommendations, trading, or – more generally – control scenarios. In the practice section, we implemented a recommender using MABs and a trading bot with a DQN.

In the next chapter, we'll look at case studies with time-series. Among other things, we'll look at multivariate forecasts of energy demand.

12

Multivariate Forecasting

As you'll have picked up by now if you've been paying attention to this book, the field of time-series has made lots of advances within the last decade. Many extensions and new techniques have popped up for applying machine learning to time-series. In each chapter, we've covered lots of different issues around forecasting, anomaly and drift detection, regression and classification, and approaches including traditional approaches, machine learning with gradient boosting and others, reinforcement learning, online learning, deep learning, and probabilistic models.

In this chapter, we'll put some of this into practice in more depth. We've covered mostly univariate time-series so far, but in this chapter, we'll go through an application of forecasting to energy demand. With ongoing energy or supply crises in different parts of the world, this is a very timely subject. We'll work with a multivariate time-series, and we'll do a multi-step forecast using different approaches.

We're going to cover the following topics:

- Forecasting a Multivariate Time-Series
- What's next for time-series?

The second section is going to cover an outlook into the future of time-series applications and research. But let's start with a discussion of multivariate series. Then we'll apply a few models to energy demand forecasting.

Forecasting a Multivariate Time-Series

Time-series forecasting is an active research topic in academia. Forecasting long-term trends is not only a fun challenge, but has important implications for strategic planning and operations research in real-world applications such as IT operations management, manufacturing, and cyber security.

A multivariate time-series has more than one dependent variable. This means that each dependent variable not only depends on its own past values, but also potentially on the past values of other variables. This introduces complexity such as colinearity, where the dependent variables are not independent, but rather correlated. Colinearity violates the assumptions of many linear models, and it is therefore even more appealing to resort to models that can capture feature interactions.

This figure shows an example of a multivariate time-series, COVID deaths in different countries (from the English Wikipedia article about the COVID-19 pandemic):

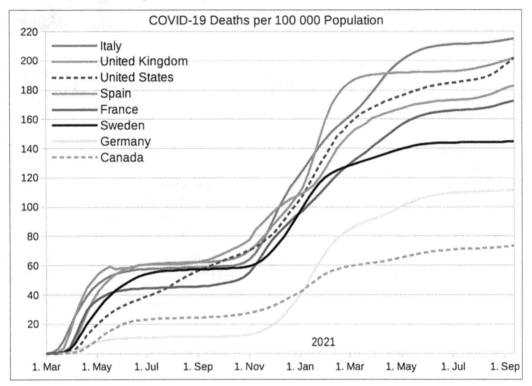

Figure 12.1: COVID-19 deaths per 100,000 population as an example of a multivariate time-series.

COVID fatalities are correlated between different countries, although they might be shifted, or they might fall into different groups.

We've mentioned the Makridakis Competitions, in *Chapter 5, Introduction to Machine Learning for Time-Series*. Spyros Makridakis, the chief organizer, is a professor at the University of Nicosia, and specializes in time-series forecasting. These competitions serve as a benchmark of the best algorithms and researchers and practitioners compete against each other for cash prices. The hope for this competition is that it can inspire and act as catalyst for machine learning, and open up directions for future work.

The M4 competition used 100,000 multivariate time-series (from the ForeDeCk database) covering different application domains and temporal scales, and results were published in 2020. 49 contestants or teams submitted point forecasts testing the accuracy of major ML and statistical methods.

The M4 organizers, Spyros Makridakis, Evangelos Spiliotis, and Vassilios Assimakopoulos, observed ("*The M4 Competition: 100,000 time-series and 61 forecasting methods*", 2020) that combinations (hybrids or ensembles) of mostly well-established statistical methods tended to be more accurate than either pure statistical or pure ML methods, which performed rather poorly, mostly placed in the second half of the field. Although there's increasing adoption of machine learning methods in solving forecasting challenges, statistical methods remain powerful, especially while dealing with low-granularity data. It should be noted, however, that the datasets didn't include exogenous variables or time-stamps. Deep learning and other machine learning methods could perhaps make better use of higher dimensionality, especially in the presence of collinearity, so this additional information would perhaps have boosted the performance of these models.

However, Slawek Smyl from Uber Technologies came in first place, taking home €9000 with a hybrid between a recurrent neural network and a statistical time-series model (Holt-Winters exponential smoothing). These two components were fit concurrently using gradient descent. A seasoned time-series practitioner, Smyl had previously won the *Computational Intelligence in Forecasting International Time-Series Competition 2016* using recurrent neural networks. It can be argued that this result shows that pragmatism with machine learning (and deep learning as an extension) can pay off.

Economists have long been working with mixtures in forecasting such as Gaussian mixture models or mixtures of GARCH models. The `Skaters` library comes with various functionality for ensembles and also does ensembles of ARMA and similar models. You can find an overview of different ensemble models on the microprediction time-series leaderboard: `https://microprediction.github.io/timeseries-elo-ratings/html_leaderboards/overall.html`

More on the machine learning side, a common method for ensembles, particularly in bagging, is training several models and weighting their predictions by their performance. Bagging uses sampling with replacement to create training samples to fit the base models. The out-of-bag (OOB) error is the mean prediction error of a model on training samples that weren't part of the training set.

Ensembles can also be composed of base models of different types, called heterogeneous ensembles. Scikit-learn provides stacking for regression and classification, where a final model can find coefficients, weighted to the base model predictions, to combine base model predictions.

There are still many pain points in industry workflows for time-series analytics. Chief among them is that there aren't many software libraries that support multivariate forecasting.

As of September 2021, although it's on the roadmap, multivariate forecasting is not part of the Kats library (even though there's support for multivariate classification). There are VAR and VARMAX models in the statsmodels library; however, there's no support for deseasonalizing multivariate time-series.

Salesforce's Merlion library claims to support multivariate forecasts, but it doesn't seem to be part of the current functionality. The Darts library provides several models that would work for multivariate forecasts.

Neural networks and ensembles such as Random Forest or boosted decision trees support being trained on multivariate time-series. In *Chapter 7*, *Machine Learning Models for Time-Series*, we worked with XGBoost to create an ensemble model for time-series forecasting. In the GitHub repository accompanying this book, I've attached a notebook that shows how scikit-learn pipelines and multioutput regressors can be applied to multivariate forecasts. In this chapter, however, we'll focus on deep learning models.

Alejandro Pasos Ruiz and colleagues at the University of East Anglia (Norwich, Norfolk, United Kingdom) highlight how multivariate applications have been neglected in their paper *"The great multivariate time-series classification bake off: a review and experimental evaluation of recent algorithmic advances"* (2020). There was a large focus on modeling univariate datasets, as is evident not only in the availability of software solutions, but also in datasets, previous competitions, and research.

They ran a benchmark of time-series classification on 30 multivariate time-series from the UEA dataset. They found that three classifiers are significantly more accurate than the dynamic time warping algorithm: HIVE-COTE, CIF, and ROCKET (please refer to *Chapter 4*, *Introduction to Machine Learning for Time-Series*, for details around these methods); however, the deep learning approach ResNet wasn't very far from these front-runners.

In the paper *"Deep learning for time-series classification: a review"* by *Hassan Ismail Fawaz* and others (2019), one of the findings from a benchmark test was that some deep neural networks can be competitive with other methods. They later followed this up by showing that neural network ensembles are on-par with HIVE-COTE on the same data (*"Deep Neural Network Ensembles for Time-Series Classification,"* 2019).

Pedro Lara-Benítez and others (2021) did another comparison in their paper *"An Experimental Review on Deep Learning Architectures for Time-Series Forecasting."* They ran an Echo State Network (ESN), a Convolutional Neural Network (CNN), a Temporal Convolutional Network (TCN), a fully connected feedforward network (MLP), and several recurrent architectures such as Elman Recurrent Networks, Gated Recurrent Unit (GRU) networks, and Long Short-Term Memory (LSTM) networks.

Statistically, based on average ranks, CNN, MLP, LSTM, TCN, GRU, and ESN were indistinguishable.

On the whole, deep learning models are very promising, and because of their flexibility they can fill the existing gap for multivariate forecasting. I hope to demonstrate in this chapter how useful they can be.

We'll be applying the following models in this chapter:

- N-BEATS
- Amazon's DeepAR
- Recurrent neural network (LSTM)
- Transformer
- Temporal convolutional network (TCN)
- Gaussian process

We went through the details for most of these methods in *Chapter 10, Deep Learning for Time-Series*, but I'll briefly cover the main features for each in turn.

Neural Basis Expansion Analysis for interpretable Time-Series forecasting (N-BEATS), presented at the ICLR conference 2020, achieved a 3% improvement over the winner of the M4 competition. The authors demonstrated that a pure deep learning approach, without any time-series-specific components, outperforms statistical approaches to challenging datasets such as the M3 and M4 competition datasets and the TOURISM dataset. A further advantage of this approach is that it is interpretable (although we won't be focusing on this aspect in the current chapter).

DeepAR is a probabilistic auto-regressive recurrent network model coming out of Amazon Research Germany. They compared the accuracy of the quantile predictions for three different datasets and only compared the forecasting accuracy against a factorization technique (MatFact) and on two datasets (traffic and electricity).

Long Short-Term Models (LSTM) networks are used for sequence modeling. A great selling point of recurrent neural networks such as LSTMs is that they can learn long-term sequences of data points.

Transformers are attention-based neural networks, originally presented in the 2017 paper *"Attention Is All You Need."* Their key features are linear complexity with the number of features and long-term memory, giving us access to any point in the sequence directly. An advantage of transformers over recurrent neural networks is that they are executed in parallel rather than in sequence and therefore run faster in both training and prediction.

Transformers were designed to solve sequences problems in **Natural Language Processing** (**NLP**) tasks; however, they can equally be applied to time-series problems, including forecasting, although such applications don't make use of features more specific to sentences, such as positional encoding.

A **Temporal Convolutional Network** (**TCN**) consists of dilated, causal, 1D convolutional layers with the same input and output lengths. We are using an implementation that includes residual blocks as proposed by Shaojie Bai and others (2018).

The last of these methods, **Gaussian processes** can't be convincingly categorized as deep learning models; however, they are equivalent to a single-layer fully-connected neural network with an independent and identically distributed prior over its parameters. They can be seen as an infinite-dimensional generalization of multivariate normal distributions.

An interesting, additional aspect – although, again, we won't pursue this here – is that many of these methods allow using additional explanatory (exogenous) variables.

We'll be using a 10-dimensional time-series of energy demand in different states. The dataset comes from the 2017 Global Energy Forecasting Competition (GEFCom2017).

Each variable records the energy usage in a particular region. This emphasizes the problems with long-term memory – to highlight this, we'll be doing a multi-step forecast.

You can find the `tensorflow/keras` implementations of the models together with utility functions for the data on GitHub in a repository I created for demonstration purposes featuring time-series models for multivariate and multi-step forecasting, regression, and classification: `https://github.com/benman1/time-series`.

Let's jump right into it.

Python practice

We'll load the dataset of the energy demand, and we'll apply several forecasting methods. We are using a big dataset and some of these models are quite complex, so training can take a long time. I would advise you to use Google Colab and switch on GPU support, or to reduce the number of iterations or the size of the dataset. I'll mention performance tweaks later when they become relevant.

Let's start by installing the library from the GitHub repository mentioned above:

```
!pip install git+https://github.com/benman1/time-series
```

This shouldn't take long. Since the requirements include `tensorflow` and `numpy`, I'd recommend installing them into a virtual environment.

Then, we'll load the dataset using a utility method in the library and wrap it in a
`TrainingDataSet` class:

```
from time_series.dataset.utils import get_energy_demand
from time_series.dataset.time_series import TrainingDataSet

train_df = get_energy_demand()
tds = TrainingDataSet(train_df)
```

If you wanted to speed up the training, you could reduce the number of
training samples. For instance, instead of the previous line, you could say: `tds =
TrainingDataSet(train_df.head(500))`.

We'll do this for the `GaussianProcess` later, which can't handle the full dataset.

For most of these models, we'll use TensorFlow graph models, which depend on
non-eager execution. We'll have to disable eager execution explicitly. Also, for one
of the models, we need to set up output of intermediates to avoid a TensorFlow
problem: `Connecting to invalid output X of source node Y which has Z
outputs`:

```
from tensorflow.python.framework.ops import disable_eager_execution
import tensorflow as tf
disable_eager_execution()  # for graph mode
tf.compat.v1.experimental.output_all_intermediates(True)
```

I've set up metrics and plotting methods that we'll use for all the produced forecasts.
We can just load them up from the time-series library:

```
from time_series.utils import evaluate_model
```

We'll also set the number of epochs in training to `100` – the same for every model:

```
N_EPOCHS = 100
```

If you find that training is taking very long, you can set this to a lower value so
training finishes earlier.

Let's go through the different forecasting methods in turn, `DeepAR` first:

```
from time_series.models.deepar import DeepAR
ar_model = DeepAR(tds)
ar_model.instantiate_and_fit(verbose=1, epochs=N_EPOCHS)
```

We'll see the summary of the model and then the training error over time (omitted here):

```
Layer (type)                Output Shape              Param #
=================================================================
input_2 (InputLayer)        [(None, 10, 10)]          0

lstm_1 (LSTM)               (None, 10, 4)             240

dense_1 (Dense)             (None, 10, 4)             20

main_output (GaussianLayer) [(None, 10, 10), (None, 1 100
=================================================================
Total params: 360
Trainable params: 360
Non-trainable params: 0
```

Figure 12.2: DeepAR model parameters.

This model is relatively simple, as we can see: only 360 parameters. Obviously, we could tweak these parameters and add more.

We'll then produce predictions on the test dataset:

```
y_predicted = ar_model.model.predict(tds.X_test)
evaluate_model(tds=tds, y_predicted=y_predicted,
    columns=train_df.columns, first_n=10)
```

We'll see the errors – first the overall error and then for each of the 10 dimensions:

```
MSE: 0.4338
----------
CT: 0.39
MASS: 1.02
ME: 1.13
NEMASSBOST: 1.48
NH: 1.65
RI: 1.48
SEMASS: 1.65
TOTAL: 1.45
VT: 1.23
WCMASS: 1.54
```

We'll see the plot over the first 10 time-steps:

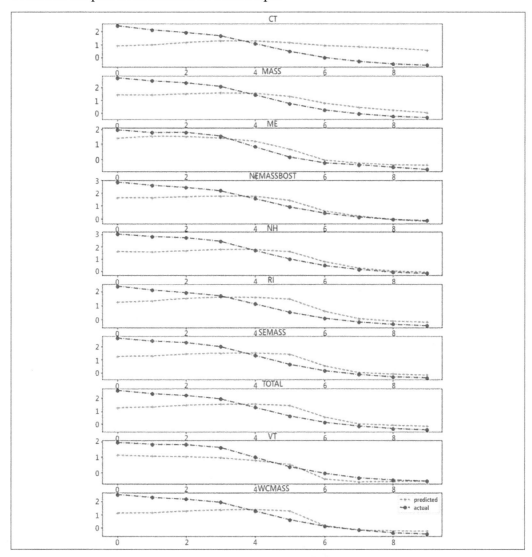

Figure 12.3: DeepAR forecasts for 10 time-steps.

Let's move on to the next method: N-BEATS:

```
from time_series.models.nbeats import NBeatsNet
nb = NBeatsNet(tds)
nb.instantiate_and_fit(verbose=1, epochs=N_EPOCHS)
y_predicted = nb.model.predict(tds.X_test)
evaluate_model(tds=tds, y_predicted=y_predicted,
    columns=train_df.columns, first_n=10)
```

N-BEATS trains two networks. The forward network has 1,217,024 parameters.

Let's see the forecasts:

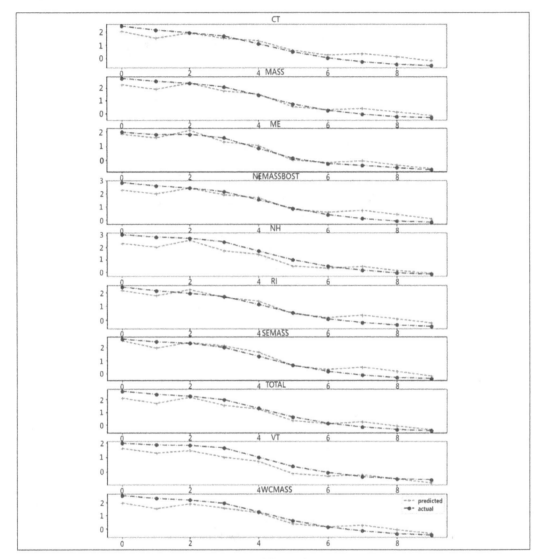

Figure 12.4: N-BEATS forecasts.

LSTM is next:

```
from time_series.models.LSTM import LSTM
lstm = LSTM(tds)
lstm.instantiate_and_fit(verbose=1, epochs=N_EPOCHS)
y_predicted = lstm.model.predict(tds.X_test)
evaluate_model(tds=tds, y_predicted=y_predicted,
    columns=train_df.columns, first_n=10)
```

This model takes a lot more parameters than DeepAR:

```
Layer (type)              Output Shape           Param #
=========================================================
input_3 (InputLayer)      [(None, 10, 10)]       0

lstm_2 (LSTM)             (None, 10, 100)        44400

dense_2 (Dense)          (None, 10, 10)         1010
=========================================================
Total params: 45,410
Trainable params: 45,410
Non-trainable params: 0
```

Figure 12.5: LSTM model parameters.

`45,000` parameters – this means this takes much longer to train than `DeepAR`.

Here we see the forecasts again:

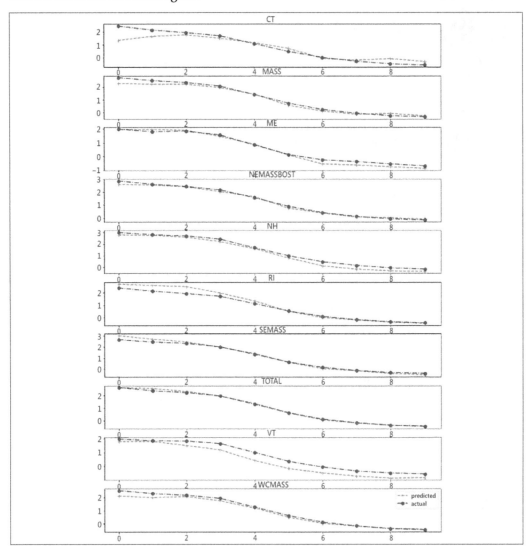

Figure 12.6: LSTM forecasts.

Let's do the transformer:

```
trans = Transformer(tds)
trans.instantiate_and_fit(verbose=1, epochs=N_EPOCHS)
y_predicted = trans.model.predict(tds.X_test)
evaluate_model(tds=tds, y_predicted=y_predicted,
    columns=train_df.columns, first_n=10)
```

Here's the forecast plot:

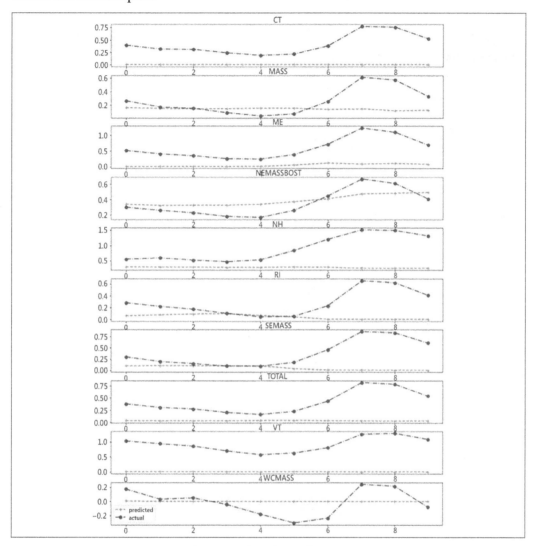

Figure 12.7: Transformer forecasts.

This model takes very long to train and the performance was the worst of the bunch.

Our last deep learning model is the TCN:

```
from time_series.models.TCN import TCNModel
tcn_model = TCNModel(tds)
tcn_model.instantiate_and_fit(verbose=1, epochs=N_EPOCHS)
print(tcn_model.model.evaluate(tds.X_test, tds.y_test))
```

```
y_predicted = tcn_model.model.predict(tds.X_test)
evaluate_model(tds=tds, y_predicted=y_predicted, columns=train_
df.columns, first_n=10
```

The forecasts are as follows:

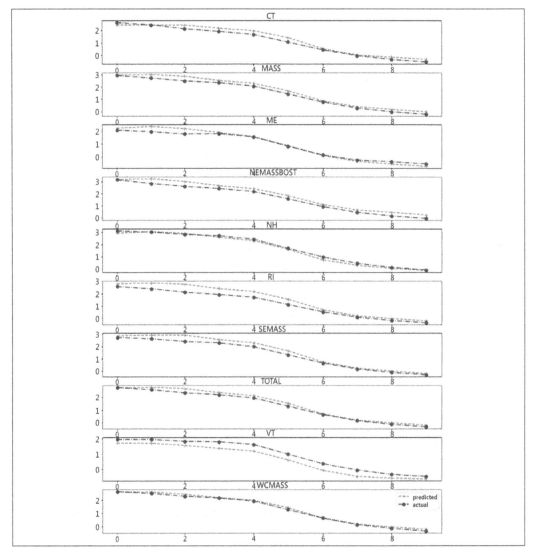

Figure 12.8: TCN forecasts.

The Gaussian process, unfortunately, can't deal with our dataset – therefore, we'll only load up a small part. The Gaussian process also depends on eager execution, so we'll have to restart the kernel, redo the imports, and then execute this. If you have doubts about how to do this, please have a look at the `gaussian_process` notebook in the GitHub repository of this book.

Here we go:

```
from time_series.models.gaussian_process import GaussianProcess
tds2d = TrainingDataSet(train_df.head(500), train_split=0.1, two_dim=True)
gp = GaussianProcess(tds2d)
gp.instantiate_and_fit(maxiter=N_EPOCHS)
y_predicted = gp.predict(tds2d.X_test)[0].numpy().reshape(-1,
tds.dimensions, tds.n_steps)
evaluate_model(tds=tds, y_predicted=y_predicted,
    columns=train_df.columns, first_n=10)
```

The forecast looks like this:

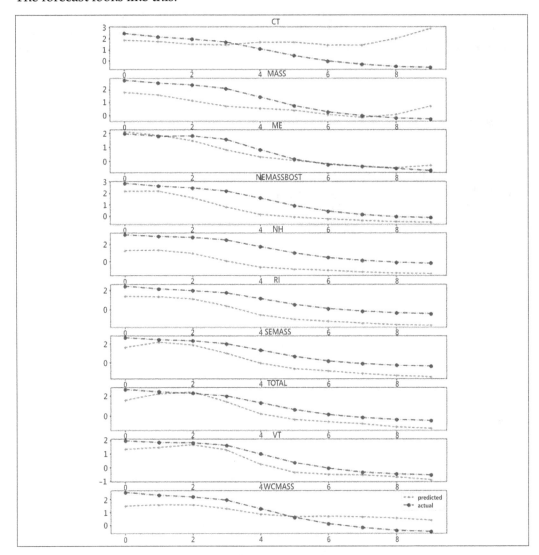

Figure 12.9: Gaussian process forecasts.

All algorithms (except for the Gaussian process) were trained on 99336 data points. As mentioned, we've set the training epochs to 100, but there's an early stopping rule that would stop training if the training loss didn't change within 5 iterations.

The models are validated on the test set.

Let's check the statistics:

	Parameters	MSE (test)	Epochs
DeepAR	360	0.4338	100
N-BEATS	1,217,024	0.1016	100
LSTM	45,410	0.1569	100
Transformer	51,702	0.9314	55
TCN	145,060	0.0638	100
Gaussian process	8	0.4221	100
ES	1	11.28	-

Given the huge disparity in the error between the deep learning methods, there might be something off with the implementation of the transformer – I'll try to fix this at some point.

I've included a baseline method, **Exponential Smoothing (ES)**, in the mix. You can find the code for this in the time-series repository.

This brings the chapter and the book to its conclusion. You can have a look at the repo if you want to understand better what's happening under the hood. You can also tweak the model parameters.

What's next for time-series?

We've looked at many aspects of time-series in this book. If you've made it this far, you should have learned how to analyze time-series, and how to apply traditional time-series forecasts. This is often the main focus of other books on the market; however, we went far beyond.

We looked at preprocessing and transformations for time-series as relevant to machine learning. We looked at many examples of applying machine learning both in an unsupervised and supervised context for forecasting and other predictions, anomaly detection, and drift and change point detection. We delved into techniques such as online learning, reinforcement learning, probabilistic models, and deep learning.

In each chapter, we've been looking at the most important libraries, sometimes even the cutting edge, and, finally, prevalent industrial applications. We've looked at state-of-the-art models such as HIVE-COTE, preprocessing methods such as ROCKET, and models that adapt to drift (adaptive online models), and we reviewed a number of methods for anomaly detection.

We've even looked at scenarios such as switching between time-series models with multi-armed bandits or causal analysis with counterfactuals.

Due to their prevalence, time-series modeling and forecasting are crucial in multiple domains and have great economic importance. While traditional and well-established approaches have been dominating, machine learning for time-series is a relatively new research field, having only really just come out of its infancy, and deep learning is a very active forefront of this revolution.

The search for good models will carry on, extending to bigger new challenges. One of these, as I hoped to show in the preceding section of this chapter, is making multivariate methods a more practical proposition.

The next Makridakis Competition, M5, focuses on hierarchical time-series provided by Walmart (42,000 time-series). Final results will be published in 2022. Machine learning models can shine at hierarchical regression on time-series, outperforming some well-established models in the literature as shown by *Mahdi Abolghasemi* and others ("*Machine learning applications in time-series hierarchical forecasting*," 2019) in a benchmark with 61 groups of time-series with different volatilities. Mixed-effects models (with application to groups and hierarchies) for time-series forecasting is also an active area of research.

The M6 competition features real-time financial forecasting of S&P500 US stocks and international ETFs. Future competitions might focus on non-linearities such as Black Swan events, time-series with fat tails, and distributions that are important for risk management and decision making.

packt.com

Subscribe to our online digital library for full access to over 7,000 books and videos, as well as industry leading tools to help you plan your personal development and advance your career. For more information, please visit our website.

Why subscribe?

- Spend less time learning and more time coding with practical eBooks and Videos from over 4,000 industry professionals
- Learn better with Skill Plans built especially for you
- Get a free eBook or video every month
- Fully searchable for easy access to vital information
- Copy and paste, print, and bookmark content

Did you know that Packt offers eBook versions of every book published, with PDF and ePub files available? You can upgrade to the eBook version at www.Packt.com and as a print book customer, you are entitled to a discount on the eBook copy. Get in touch with us at customercare@packtpub.com for more details.

At www.Packt.com, you can also read a collection of free technical articles, sign up for a range of free newsletters, and receive exclusive discounts and offers on Packt books and eBooks.

Other Books You May Enjoy

If you enjoyed this book, you may be interested in these other books by Packt:

Learn Python Programming – Third Edition

Fabrizio Romano

Heinrich Kruger

ISBN: 978-1-80181-509-3

- Get Python up and running on Windows, Mac, and Linux

- Write elegant, reusable, and efficient code in any situation

- Avoid common pitfalls like duplication, complicated design, and over-engineering

- Understand when to use the functional or object-oriented approach to programming

- Build a simple API with FastAPI and program GUI applications with Tkinter

- Get an initial overview of more complex topics such as data persistence and cryptography

- Fetch, clean, and manipulate data, making efficient use of Python's built-in data structures

Python Object-Oriented Programming – Fourth Edition

Steven F. Lott

Dusty Phillips

ISBN: 978-1-80107-726-2

- Implement objects in Python by creating classes and defining methods

- Extend class functionality using inheritance

- Use exceptions to handle unusual situations cleanly

- Understand when to use object-oriented features, and more importantly, when not to use them

- Discover several widely used design patterns and how they are implemented in Python

- Uncover the simplicity of unit and integration testing and understand why they are so important

- Learn to statically type check your dynamic code

- Understand concurrency with asyncio and how it speeds up programs

Expert Python Programming – Fourth Edition

Michał Jaworski

Tarek Ziadé

ISBN: 978-1-80107-110-9

- Explore modern ways of setting up repeatable and consistent Python development environments
- Effectively package Python code for community and production use
- Learn modern syntax elements of Python programming, such as f-strings, enums, and lambda functions
- Demystify metaprogramming in Python with metaclasses
- Write concurrent code in Python
- Extend and integrate Python with code written in C and C++

Packt is searching for authors like you

If you're interested in becoming an author for Packt, please visit authors.packtpub.com and apply today. We have worked with thousands of developers and tech professionals, just like you, to help them share their insight with the global tech community. You can make a general application, apply for a specific hot topic that we are recruiting an author for, or submit your own idea.

Share your thoughts

Now you've finished *Machine Learning for Time-Series with Python*, we'd love to hear your thoughts! Scan the QR code below to go straight to the Amazon review page for this book and share your feedback or leave a review on the site that you purchased it from.

https://packt.link/r/1801819629

Your review is important to us and the tech community and will help us make sure we're delivering excellent quality content.

Index

standard deviation 44
standard error (SE) 45
stationarity 6, 56, 136, 137
stationary processes 56, 136
Statsmodels 146, 147
Statsmodels library
  using, for modeling 147
Structured Query Language (SQL) 30
style guide for PEP 8
  reference link 32
sunlight hours
  obtaining, for specific day 87
supervised algorithms, for regression and
     classification
  implementations 128, 129
supervised learning 96, 97
Support Vector Machines (SVMs) 103, 271
Suspended Particulate Matter (SPM) 55
Symbolic Aggregate ApproXimation
     (SAX) 121
Symbolic Fourier Approximation (SFA) 122
symmetric mean absolute percentage
     error (SMAPE) 111
synapses 263

T

Temporal Convolutional
     Network (TCN) 276, 322, 331-333
Temporal Dictionary Ensemble (TDE) 124
Temporal Difference (TD) learning 299
Temporal Fusion Transformer (TFT) 278
Theil's U 113
Theta method 141
Thompson sampling 303
time-series 3, 6, 335, 336
  characteristics 4, 5
  comparing 116
  machine learning methods, using 186, 187
  offline learning 210
  online learning 210
  reference link 324
  reinforcement learning (RL) 301
  unsupervised methods 162-164
  working with, in Python 38
time-series analysis (TSA) 6, 36-38

time-series classification algorithms
  critical difference diagram 126, 127
Time-Series Combination of Heterogeneous
     and Integrated Embedding
     Forest (TS-CHIEF) 121
time-series data
  examples 2
time-series datasets 94
time-series forecasting 97
Time-Series Forest (TSF) 120
time-series machine learning algorithms
  detailed taxonomy 125, 126
time-series machine learning flywheel 38
time-series regression 107
transformer 330, 331
transformer architectures 277
trend 56
  identifying 56-63
triple exponential smoothing 142
true positive rate (TPR) 114, 115
true positives (TP) 115
Twitter 170

U

UAE (University of East Anglia) 118
UCR (University of California, Riverside) 118
unit imputation 73, 82
univariate analysis 38
univariate series 4
Universidade Federal de Minas
     Gerais (UFMG) 252
unsupervised learning 96, 97
unsupervised methods
  for time-series 162-164

V

validation 187, 188
value-based learning 300
variables 44-49
  relationships 49, 50, 52
vector autoregression models 144, 145
Vector Autoregressions (VAR) 129, 133
Very Fast Decision Tree (VFDT) 215
virtual environment 193

W

walk-forward validation 188
weak learner 100
WEASEL+MUSE 123
window-based features 75
Wold's decomposition 137
Word Extraction for Time-Series
        Classification 123

Y

Yeo-Johnson transformation 72

Z

Z-score normalization 70